Ravaged Relentlessly

Biblical Help for Those Overcoming Sexual Crime

Kesha Lee Gore

AuthorHouse™
1663 Liberty Drive, Suite 200
Bloomington, IN 47403
www.authorhouse.com
Phone: 1-800-839-8640

First published by AuthorHouse 1/23/2008

ISBN: 978-1-4343-3901-0 (sc)

Printed in the United States of America
Bloomington, Indiana

This book is printed on acid-free paper.

Unless otherwise noted, all scripture quotations in this book are from the
King James Version of the Bible.

SINCERE Ministries, Inc.
P.O. Box 54001
Washington, D.C. 20032
www.sinceremininistries.org

Published by SINCERE Ministries, Inc.
&
Authorhouse, (Author Solutions, Inc.)

Printed in U.S.A.

Dedication:

To You Lord, I have to say thank you for teaching me. This has been a wonderful experience, I have learned so much, you are amazing. Never in a lifetime could I have ever accomplished such a compilation of wisdom and truth and be of help to someone else because of it. You are truly all wise, without you, I know absolutely nothing. Thank you for trusting me to listen to you and record this work for the benefit of us all.

"…and a certain woman which had an issue… suffered many things of many physicians, and had spent all that she had, and was nothing bettered, but rather grew worse; When she had heard of Jesus, came in the press behind, and touched his garment. For she said, If I may touch but his clothes, I shall be whole. And straightway…she felt…that she was healed of that plague. And Jesus, immediately knowing in himself that virtue had gone out of him, turned about in the press, and said, Who touched my clothes? …But the woman fearing and trembling, knowing what was done in her, came and fell down before him, and told him all the truth. And he said unto her, Daughter, thy faith hath made thee whole; go in peace, and be whole…" Mark 5:25-34

Table of Contents

Preface

There is a passion in me to help people just like you to understand the heart of God towards victims of sexual crime. I know a truth that is powerful enough to make right every wrong that you have known thus far. Originally, I sought out answers for myself, that is, for my own problems; but in my searching I found out that I was not alone. Once I was blessed to know the answer to my heart's cry, I was compelled to share that truth with the following individuals:

- The apparent 84% of rape victims who don't report their experience.
- The Asian 'Comfort Women' taken by the Japanese in WWII, who had to be medicated to endure the pain of being physically used by men daily.
- The over 4,500 women in the Eastern Congo that were raped and gang-raped this year, (2007), by the militia groups that use sexual assault as an act of terrorism; many of them public and some in front of their husbands.
- The wives of the ones who changed their lives when they assaulted them, refused to let them go, then insisted on marrying them.
- The children who don't know what a good night's sleep is.
- The boys who fearfully live threatened if they tell their story.
- The soldiers, cadets and those in the care of the Military who were sexually violated but found no justice because the Military valued its honor over that of the victim.
- Those brave souls who reported the crimes against them and were not believed.

- The grieving man who won't tell a soul the real reason he isn't married.
- The ones who do marry and are never able to fully connect with their spouses, even though they love them dearly.
- The women, girls and boys who were kidnapped and made into sex slaves.
- The boys and girls who acted out their sorrows by turning to prostitution.
- The prostitutes who feel that the streets hurt less than home.
- The ones who helplessly watch their loved one suffer because of sexual crime day after day.
- The women, who are ravaged relentlessly as punishment in the name of religious honor.
- The precious souls whose hearts have been broken from sexual crime, by those that they trusted with their lives.
- The ones who live in a prison called home where God only knows what they are subjected to.
- The prisoners who are physical targets of those struggling for power.
- The ones who live outrageous and promiscuous lives as if they are disposable, to try to redeem their stolen power back.
- The boys who were always embraced by females and rejected by males –thereby never coming to know who they really are.
- The girls who felt like they were invisible compared to other females.
- The elderly who, for decades, have secretly carried the pain of being sexually humiliated.
- The elderly who want to make right the wrong they did to others as a result of what was done to them.
- The ones who lost their ability to speak again, dance again,

stroll again, jog again, sing again, trust again– live to the fullest again.

- The runaways who ran into even more danger and pain.
- The children stuck in foster homes where they are abused.
- The one who ran away from a job, a hobby, a personality, a lifestyle, a pastime, a city, a friendship, an identity, a marriage – even a country.
- The ones who feel like God abandoned them.
- The ones who abandoned their real selves.
- The ones who blame their pain on God.
- The ones who blame themselves.
- The ones who can't bring themselves to think about it much less talk about it.
- The ones who would be cured of the disease the assailant left behind with its' symptoms of hatred, powerlessness, fear, insecurity, anger, anxiety and confusion.
- The one who wants to know what God has to say about all of this.
- Those who wonder if anyone in the Bible ever went through this struggle and if so, did they recover?
- Those who have a million questions they will never ask out loud.

The list can go on and on, these are just examples of what my studying has revealed to me. There are, I am sure, more instances and episodes than I dare even imagine, but this work has little to do with the terrible things that happen to us as a people. To write only about this dark issue would be redundant, because we all have been touched by sexual crime directly, indirectly or historically. This message is to help you come back to yourself after surviving that terrible thing, no matter when it happened, how awful it was or who subjected you to it.

In all my seeking and frustration, I found other victims, but no answer; until the day I sought the Lord. Let me just tell you, no greater peace have I found than in the comforting embrace of God's own words. My entire life and outlook has been changed ever since. I have written what He taught me in this book as honestly and transparently as I could. Let me share this hope with you.

Introduction:

The first time I saw the word ravished was in a text in the Bible. It was in the context of a pronounced curse on a wicked and disobedient people. It reads, *"They ravished the women in Zion, and the maids in the cities of Judah."* *Lamentations 5:11*

When I saw it, I knew what it meant because of how it was written. I immediately looked it up in the dictionary and found it to mean: 'to devastate; ruin; rape. To ravage.' I decided to use the word ravage to title this book because it covers more than just rape. It encircles molestation, incest, sexual abuse, sexual assault, emasculation, -- the whole scope of offense in this area.

At times a person is not raped – but sexually assaulted, yet, the devastation is still great. A trust is broken. Confidence is shattered, security is compromised and peace is often lost for years.

I first found the word rape to mean 'to force sexual intercourse upon.' I also read it defined as 'any sexual act against one's will'. When I was nine years old I miraculously escaped a group of teenage boys who approached me and began to assault me. I thought I had been raped, because the dictionary stated that any sexual act against one's will is rape. In all actuality, rape was not the word I needed, I was not raped- I was ravaged relentlessly. Those boys had no regard or consideration for the horror and weakness I lived with afterwards. I was left invisibly wounded and more vulnerable to future predators later on, not knowing for years why attacks and attempts from others seemed to keep occurring up until I was 20 years old.

Inside, my spirit was weakened. My love of life and trust in people as a whole was mercilessly halted. My feelings about myself were changed and my understanding was terribly warped. My youth was almost completely taken from me. A seed had been planted in my soul, and I wondered why people were so cruel. Life after death takes on a whole new meaning when the death occurs in the realm of the soul. But thank God I didn't stay that way. God completely restored me and healed every hurt place inside of me. I write this book to let every soul know just how possible it is to be whole again.

How can I write this? My heart is so full I can scarcely pour it out on this paper. I want to say to each person reading this that there is hope. I don't want to do what has already been unsuccessfully done. I don't want to offer commiserating words that offer no life. How do you explain to a victim of sexual crime that they can live again…truly live again? Well, with the help of the very Spirit of the Lord, I will in this book.

How many times have we heard the 'I know how you feel… 'statement made to us when we are in pain? No mere man is capable of knowing that (not even by experience). Whereas one may feel hurt from a bad experience, another may feel anger or hatred from the same type of experience. Others shock. Still others endure fear or despair. The emotional toll varies with each individual. Only the One who made us individually knows how we each individually feel. (See Psalm 139:1-4)

God has a special place in His heart for all of those who have been hurt this way. Many victims turn against God after they have been ravaged relentlessly, but only because the devil subtly deceives them and they just can't see the truth.

After years of searching for answers for my own pain, to no avail, I finally found out what God's position was. He loved me and did everything necessary to help me. He started me on a pathway to peace and hope. It started when a fellow neighbor and classmate in the Bible school that I attended, ministered a word of healing to me by the Spirit of God. He didn't even know me that well. He allowed God to use him to help me and now, I am doing the same to help all of you. It is the obedience of Mr. B. Case and his heart-filled prayer that has helped me get to this place today. I know that God is going to bless him and his ministry. I pray that this book will be as effective for you as his compassion was for me.

Sexual crime and the pain it causes is only one of the horrors of this life. The Lord is willing to save us from this and all of the terrible things with which this world is contaminated. Though it is easy to say 'Why didn't the Lord protect?' it is more accurate to consider that there may have been dozens of attacks that He did protect us from by preventing them altogether. Dangers we aren't even aware of surround us, and God intervenes more than we know. God has done so much for us. He sent His only son Jesus to save us. A Savior is exactly what a victim needs.

The truth is, this world is laden with many evils, sickness, death, poverty, hate. Sexual crime is only one. Until the Lord revamps this entire world to His intended purpose, evil is just a temporary reality. The eternal reality is, God is sovereign. God can undo, re-do, and destroy whatever the enemy of our souls has done to us. This book is a tool to help bring understanding of this truth.

Each of the chapters of this book is based on the scripture that explains the specific power that God gave His Son, Jesus Christ, to help us. This is what it says:

"The Spirit of the Lord has sent me, because He has anointed me to heal the brokenhearted, to preach deliverance to the captives and recovering of sight to the blind; to set at liberty them that are bruised; to preach the Lord's favor and the day of vengeance of our God; to comfort all who mourn and to provide for those who grieve." *Isaiah 61:1-2*

This verse in the Bible is very comforting because it shows us what God's power can do for us. Any victim of sexual crime needs the broken-ness in their hearts to be healed and only Jesus Christ can do it. They need to hear the words of God for themselves so that they won't be held captive in the dark and confusing arena of pain that their attackers left them in. They need the Lord to open their eyes to see the great future that He has planned for the rest of their lives. Their lives are not over because of the crime they endured. And for those who are bruised in the deepest part of their soul, only the touch of the Lord can set you so completely free. You can do whatever you want, go where ever you want and be whoever you desire to be without bondage to fear, sorrow and dread.

Our text goes on to tell us plainly that God sincerely does favor us, He's strong enough to avenge us and is at the same time gentle enough to comfort all of those who are overcome with sorrow. God alone, not time, can end the grief once and for all. All because of the mercy provided by the Lord through Jesus His dear son. If you read that passage again, you will see yourself in it somewhere, because the enemy of our soul has harmed us all at some time. God is here for all of us. He loves us. I believe that you will understand that better when you read this book. This is what I pray for constantly.

The sections of the book are arranged in reverse order of this passage of scripture, each with three chapters that focus on the six aspects of which it speaks. The book is written as an application of the text as it relates to the state of victims of sexual crime. The first section, for example, is designed to begin the process of offering the Lord's comfort to those who are hurt from this type of crime. It will help victims to respond to the Lord's open arms, face themselves honestly and understand the consoling truth of what Jesus' work on the cross affords them. It will also highlight the story of Jacob's daughter Dinah throughout the section.

The next five sections will likewise focus on the main text to bring understanding of God's vengeance, liberty, ability to grant sight to those who can no longer see with the eyes of their heart, the power of His deliverance and the greatness of being healed inside by His gentle touch. Each section will follow suit and highlight several characters in the Bible that were recorded as having various experiences. This includes being ravaged by authoritative figures, war induced rape, incest/rape, statutory rape, male on male rape, trick rape by women via intoxication, sodomy/gang rape, conception because of rape and one who ended up marrying (against her will and control), the one that raped her.

Each chapter will include added notes for family, loved ones and those who know the victim. This will help them to better understand the victim's struggle so that they can be a better-informed vessel of consolation and support.

This book is not the answer, but it points to the One who is. It will address every age from the very young to the very old, both male and female. Ravaged Relentlessly is not another sad story, it is a way to show every reader glimpses of the Lord's heart. Only He can change our sad stories around and make everything work out for the best in the end.

The last section of the book has special prayers for the victims and their loved ones that were individually written by some of the Associates of SINCERE Ministries. Truly our prayers are with you all. You will also find throughout the text several scripture passages highlighted for you to learn and meditate on as you embrace God's healing and deliverance. There are even journal pages for you to record how the Lord has worked in your heart as you read through the book. In the future you may refer to them when you share of what the Lord did in your life (I just know that when you read this book, you will open your heart to let Him bless you). Your testimony will help others just as the examples in this book will help you.

So here it is. A key to help you unlock the door of your heart to receive the Lord's helping hand. He wants to embrace every one of you and bless you with a new life. He loves you, even though the devil may have been whispering to you that He does not. Nothing could be further from the truth; God loves you and cares for you. He will do anything for you. He's even coming again one day to once and for all end the evil work of that old devil, the one who is behind all the hurt that we go through. Never again will we have to cry or be grieved, never. So read on. God has something to say to all those, and those who care for those who have been…

Ravaged Relentlessly

section one:

*"To **comfort** all who mourn and to provide for those who grieve…"*

"Dinah the daughter of Leah, which she bare unto Jacob, went out to see the daughters of the land. And when Shechem the son of Hamor the Hivite, prince of the country, saw her, he took her and lay with her, and defiled her. And his soul clave unto Dinah the daughter of Jacob, and he loved the damsel, and spake kindly unto the damsel. And Shechem spake unto his father Hamor, saying, 'Get me this damsel to wife.'

And Jacob heard that he had defiled Dinah his daughter: now his sons were with his cattle in the field: and Jacob held his peace until they were come. And Hamor the father of Shechem went out unto Jacob to commune with him. And the sons of Jacob came out of the field when they heard it: and the men were very wroth, because he had wrought folly in Israel in lying with Jacob's daughter; which thing ought not to be done. And Hamor communed with them, saying, 'The soul of my son Shechem longeth for your daughter: I pray you give her him to wife. And make ye marriages with us, and give your daughters unto us, and take our daughters unto you. And ye shall dwell with us: and the land shall be before you; dwell and trade ye therein, and get you possessions therein.' And Shechem said unto her father and brethren, 'Let me find grace in your eyes, and what ye shall say unto me I will give. Ask me never so much dowry and gift and I will give according as ye shall say unto me: but give me the damsel to wife.' And the sons of Jacob answered Shechem and Hamor his father deceitfully, and said, because he had defiled Dinah their sister: And they said unto them, 'we cannot do this thing, to give our sister to one that is uncircumcised; for that were a reproach unto us: But in this will we consent unto you: If ye will be as we be, that every male of you be circumcised; then will we give our daughters unto you and we will take your daughters to us, and we will dwell with you, and we will become one people. But if ye will not hearken unto us, to be circumcised; then will we take our daughter, and we will be gone'. And their words pleased Hamor, and Shechem Hamor's son.

And the young man deferred not to do the thing, because he had delight in Jacob's daughter: and he was more honourable than all the house of his father. And Hamor and Shechem his son came unto the gate of their city, and communed with the men of their city, saying, 'these men are peaceable with us; therefore let him dwell in the land, and trade therein; for the land, behold, it is large enough for them; let us take their daughters to us for wives, and let us give them our daughters. Only herein will the men consent unto us for to dwell with us, to be one people, if every male among us be circumcised as they are circumcised. Shall not their cattle and their substance and every beast of their's be ours? Only let us consent unto them, and they will dwell with us.' And unto Hamor and unto Shechem his son hearkened all that went out of the gate of his city; and every male was circumcised, all that went of the gate of his city.

And it came to pass on the third day, when they were sore, that two of the sons of Jacob, Simeon and Levi, Dinah's brethren, took each man his sword, and came upon the city boldly, and slew all the males. And they slew Hamor and Shechem his son with the edge of the sword, and took Dinah out of Shechem's house, and went out. The sons of Jacob came upon the slain, and spoiled the city, because they had defiled their sister. They took their sheep, and their oxen, and their asses, and that which was in the field, and all their wealth, and all their little ones, and their wives took they captive, and spoiled even all that was in the house.

And Jacob said to Simeon and Levi, 'Ye have troubled me to make me to stink among the inhabitants of the land, among the Canaanites and the Perizites: and I being few in number, they shall gather themselves together against me, and slay me; and I shall be destroyed, I and my house.' And they said, 'Should he deal with our sister as with an harlot?'

"And God said unto Jacob, 'Arise, go up to Bethel, and dwell there: and make there an altar unto God that appeared unto thee when thou fleddest from the face of Esau thy brother. ...'and they journeyed: and the terror of God was upon the cities that were round about them, and they did not pursue after the sons of Jacob.'"

Genesis 34-35: 1,5

Chapter One: Who Did This to You?

It seemed good to me to begin this book with the story of Dinah. Her story is one of the first recorded sexual assaults in the Bible and it bears great significance. Her innocence was regarded in much the same way as that of many people now. She was only about fourteen or fifteen years old, the only girl in a family of twelve boys. When her family moved to a new country, she did what any other young lady would do, she went to find other girls to befriend. Unfortunately, before she even had a chance to meet anyone, she was raped. According to the account recorded, it does not seem that it was a violent assault or forced, but more, perhaps, a coercion by the man that took advantage of her innocence. In other words, statutory rape, as we call it today is apparently what may have taken place.

When her family heard about what happened, they were deeply distressed about it. All they wanted to know, was who was responsible for the crime against their daughter and sister. The emotional reaction they had then was really no different than how loved ones respond now when a dear one has been defiled in this way. The pain is the same.

Dinah's is not a sad story. As I stated before, it has great significance. If we look at the fact that it was recorded in the Bible at all, we get the inclination that God wants to show us how He feels about rape. We can clearly see that the Lord wants us to know that it is not His desire for anyone to be hurt in this way.

When we look at the entire account of Dinah, we find that her brothers took vengeance on her behalf and rescued her out of Shechem's house, because he kidnapped her as well. There was much blood shed and Dinah's father Jacob was grieved that his sons had acted so violently. I cannot imagine what Jacob's thoughts were before his sons became vigilantes. The story does, however, express his concern for the safety of the family afterwards.

That is when the Lord brought the chaos to an end. He sent the entire family to a safe place and put fear in the hearts of everyone else, so that no one would try to avenge the lives taken by Jacob's sons. God kept them alive.

Curiosity may make us wonder, why didn't God step in before so that all of their misery could have been avoided? Here was the first daughter of the children of Israel, the Lord's chosen people, a princess, no less, and yet God didn't prevent her attacker and kidnapper from disgracing her, or keep her brothers from shedding blood. Moreover, her whole family had to be uprooted from the land they had just purchased, as a result of all of this… what was really behind all of this?

Shechem, if you recall, came from a very vile family who was far worse than he was. It is very possible that Jacob's family, being few in number, would have been overtaken by that wicked bunch in an unrelated ambush and wiped out. We know that by the father's obsession with Jacob's wealth, that he wanted an opportunity to possess it. The issue with Dinah actually gave her family an advantage, because the emotions on both sides changed the normal rules of engagement when it comes to war. Both sides used deceptive, low blows to try to overcome the other instead of initial combat. All in all, the covenant family, that is Jacob's tribe, prevailed.

Here is where we need understanding. In the larger picture of life here on earth, the human body is not the most significant. It is the soul that matters. God did step in, He kept the girl alive and He allowed her brothers to rescue her from being in a marriage (against her will), that would have, eventually, crushed her soul. She was indeed hurt, but she was not destroyed. Although this world is full of hurt and pain, the Lord does not allow the enemy to destroy our souls, even if we feel like it at times.

By the way, when the word *enemy* is used in this book, it is referring to the one the Bible calls Satan, who is the devil. He is an individual that turned against the Lord and desired to overthrow God's throne. He used to worship God as an angel, but because of his pride he challenged God. As a result, he was banished from heaven to walk the dry empty places until the time for his appointed punishment. Ever since then he has attacked, harmed, stolen from and hurt every person he can get to on his way to his eternal demise. Jesus described his behavior this way:

"The thief cometh not, but for to steal, and to kill, and to destroy: I am come that they might have life, and that they might have it more abundantly." *John 10:10*

As you see here Jesus is the one who wants to bless you, the devil, (the thief), is the one who is out to hurt you.

More personally, loved ones throughout the world and across time will always want to know 'Who did this?' to their dear ones. They will always want to know who is responsible for the pain. They will always want vengeance. They will most certainly seek justice or at the very least peace. This is the natural response. The important thing to understand is who really did this to you. Satan is the real ravager. He is the one who hurts people and he often uses humans to do it.

Why did the enemy hurt you? It isn't anything personal; it's just that he hates God. He is angry because he can never be God. See, the devil's attacks on us are aimed at opposing the will of God. In Dinah's story, Shechem's father, Hamor, wanted Dinah's father Jacob to let her marry his son because he secretly wanted to take over Jacob's wealth. Hamor's son had raped a young girl and all he could think of was getting rich. It didn't bother him that his son, who was the prince of the land, was also a rapist and a kidnapper; Hamor wanted to get to Jacob's wealth and power. Besides, the Bible says that Shechem 'was more honourable than all the house of his father.'

The devil's motive is the same in all of his attacks. He wants to get at our heavenly Father just as Hamor wanted to get at Dinah's father Jacob. The devil opposes the Lord and all that God values, including each of us. If Satan could, he would beat us down with so much sorrow and shame that we would feel worthless. But we are not worthless. We are more precious to the Lord than the most valuable of jewels. The Lord truly treasures us. God allows us to be in His treasury of our own free will.

The enemy tries to mar and damage us so that we won't want to come to the Lord. But God wants us no matter if we are scratched, scraped, marred, dulled, chipped or cracked on the surface. The Lord can easily buff out any damage the devil has done to us. We are always valuable to Him. No jeweler would throw out a diamond just because it was damaged. A diamond is a diamond. To God, we matter so much more. Sure the enemy may try to steal our joy, damage our sense of well being or destroy our hope, but the Lord will never let him get away with it. Our souls still live in spite of the enemy's advances. We can still make the choice to give ourselves to the Lord God.

Don't be fooled by the deception of the devil. Try to understand, the Lord neither hurt you nor neglected to protect you. Ever since the devil has been here on this earth, he has caused trouble. He tempted the man that God created to manage this world into turning over his rights and privileges to him. The man, named Adam, did so and now the world suffers perpetually. In other words, God gave dominion over the earth to mankind – mankind turned it over to Satan –and Satan has tried to make it miserable for mankind ever since. Sickness, poverty, pain, crime, sorrow and death all came about after that hostile takeover. In order for it to take place, Adam had to willfully disobey God. God then had to change the earth since it was under new dominion, to limit the devil's abilities with it.

The great thing is, God didn't leave mankind alone to deal with this cruel enemy. He immediately promised to provide a deliverer to save man from the clutches of the devil. That is the promise of the Christ, that is, Jesus, who is the Savior of the world. It is vital that we understand all of this in order to grasp the fact that when a person is assaulted, it really is so much more than sex- -it is a belligerent power struggle.

Now back to our main story. The devil used Shechem to attack Dinah as a way to get to Jacob (who was God's servant). Do you see how each of these characters was set up by Satan as avenues, in an attempt to defy God? Prior to all of this, the Lord had promised to bless Jacob and his family. For that blessing to get into the hands of someone else (Hamor) would be an assault to God's will.

The enemy knew Jacob would never willingly give Dinah to Shechem because it would violate Jacob's agreement or covenant, with God. So the enemy drove Shechem to rape Dinah, leaving her (at that time in history), undesirable for marriage to anyone else. Marriage to Shechem was the only apparent alternative to prevent her from being alone for the rest of her life in a state of desolation. However, marriage with an unbelieving nation would violate Jacob's covenant with God and thus halt, again, God's plan for mankind- -just as it happened with Adam. As stated before, this was not just about sex.

Is there any real difference in your situation now? No. The devil used some wretch to defile you to keep you away from coming to your promised Savior Jesus, who is God's son. This is, again, an assault to God's plan. When rape occurs, the trick is to make it look like God doesn't care or worse, does not exist at all. It is to turn you from true destiny and stay under his control (just like Shechem planned to do with Dinah). Don't listen to the lie. The devil is really the one responsible for your assault. Don't let him deceive you into staying in a state of anger, hatred and fear, never coming to be with the Father who adores you.

The enemy attacks man because man is created in God's image. He attacks woman because the seed of the woman is where the promised Savior's human lineage extends. He attacks children because they are alive unto God, innocent in spirit and full of the kind of faith that pleases and honors the Lord.

Every opportunity he gets, the devil will orchestrate attacks on people now just as he did Dinah's rape then. The surrounding circumstances may vary, but the strategy and the goal are the same. All that is needed is access. It doesn't matter the gender, age or race of the person, for all are valuable to God. Just access, isolation, some level of weakness and lack of protection. This can be in a very natural sense or spiritual.

What really happened is the enemy is the one who did this horrendous thing to you. God never hurt you at all. As soon as you stop believing the lies of the enemy and call on the Lord, He will rescue you in that very moment no matter how bad off you are. He stands ready to help you while at the same time preserving your life.

God, who is our Heavenly Father, is no more at fault when someone is raped than Dinah's father when she was. Jacob didn't rape Dinah, he took her away from the rapist to a safe place to live. He refused to give his daughter over to Shechem in marriage. He wouldn't even respond to the man's insolent request. What loving father would give his child to a rapist and kidnapper who had already caused so much grief?

Try to see God in the same way. He wants to get you away from the one who hurt you. He doesn't want you to stay in a place of pain. He wants you safe and in a place where you are truly loved. He wants to put fear in the hearts of those who would even dare think of hurting you again so that no one would attempt to touch you. God truly and honestly loves you. Look at this verse recording Jesus' words:

"...In the world ye shall have tribulation. But be of good cheer, I have overcome the world." John 16:33b

This world may indeed cause great pain. But for every attack, God is prepared to meet it. God is the Healer to all who the devil will make sick. God is the Deliverer to all those that Satan binds. God is the Redeemer of all those that the devil stole away from Him. God is the Comforter to all of those the enemy causes to grieve. He is the Avenger to those that the serpent has violated. The Lord even causes every situation to work out for the good in the end. Therefore, He is our hope.

All we must know and remember, is that when the devil does try to hurt and abduct us, God is ready to bring us back as soon as we merely breathe His name. Satan may persistently come to steal, kill and destroy all whom God loves, but he will not succeed. The Lord loves us too much to let the enemy of our souls do whatever he sets out to do. If He didn't, the devil would have destroyed all of us ages ago.

Now you know the truth of who really did this to you. I pray you also know, more importantly, who did not. God is the one who loves you. When Jesus, the Son of God and Savior of our souls, came to the earth he demonstrated the love of God in many ways and His actions prove how God feels about us. Once Jesus healed a man that was leprous after he first reached out to touch him in love and acceptance when no one else would. Another time Jesus comforted a woman who was condemned for a shameful crime she committed and He forgave her. He even raised people from the dead to comfort the grieving hearts of their loved ones. The scripture describes his lifestyle here as follows:

"...God anointed Jesus of Nazareth with the Holy Ghost and with power: who went about doing good, and healing all that were oppressed of the devil..." *Acts 10:38*

What more evidence do we need? Well if all that isn't enough, He gave his very *"...life as a ransom for many..."* Matthew 20:28. Believe me precious one, you are loved. Who else would give up everything to save us from the hold of the ravager? God simply adores us. He adores you. Please don't let what happened to you turn you away from the arms of love that God extends to you.

He wants you to come to Him and let Him love all your hurt away. You will find in Him true consolation and peace from the trouble you have known. God knows who committed this horrible act of disgrace against you. That wretch is the one to be ashamed, not you. So be comforted. God will make it all better.

Chapter Two: A Silent Conversation

If you and I could sit down together and talk about what we have been through, you probably wouldn't want to tell me about every detail of what happened to you. Maybe you've already told a few or even a lot of people, but if you are like me, you've never told a soul the whole story. Some feelings have no words to describe them. Shame blocks more than we willingly recount. I wonder what thoughts went through Dinah's mind in the space of time that passed before her brothers came and recovered her?

But if you were to talk to me, I would encourage you to say whatever was on your mind and ask me anything you wanted to. Then I would tell you exactly where in the bible the Lord addressed your situation and offer you the most earnest wisdom that I've learned.

Since we, that is, you and I, are not sitting together, we're going to have to have a silent conversation, so to speak, through this chapter. I'm going to share with you some of the thoughts and questions that I battled with after my experiences, as well as questions that I have heard other victims verbalize. (My hope is that you will be motivated to be honest with yourself about what you really feel, so that you can be comfortable enough to seek the Lord for His help).

Then I'll share the answers that the Lord revealed to me in His word by His Holy Spirit. Because of the Lord, I have recovered completely. I believe you will too. Let's begin.

Question: 'Why me?'

Answer:

"*...because your adversary, the devil, as a roaring lion walketh about, seeking whom he may devour...*" *I Peter 5:8*

The enemy of God worked the evil act that was done to us. Because God loves us, the devil attacks us to grieve Him. It is nothing personal about you or me, we just happen to be precious to the Lord.

It's not your personality, your demeanor, your style of clothes, your size, your age, your gender, your looks or any one particular thing that encouraged the attack. It was an area of weakness that the enemy preyed on. That weakness may simply have been that you were alone. Remember Dinah? She was alone in a strange new land when she was attacked. Being the daughter of Jacob, the chosen man of God, the enemy was just looking for a chance to ruin her and cause Jacob to turn against God in anger and hurt. This would thus cause him to break his covenant with God and quell our hope.

As for me, as the Lord later revealed to me, the part of me that the enemy regarded as weakness, was my quiet nature and my solitary life. Those who harmed me never believed that I would have the nerve to report them. In two out of the three times, their assumptions were right.

The first police officer to investigate my allegation asked me if I screamed. When I told him 'no', he frowned. Then he asked if I fought back. When I answered 'I tried to…', he started yelling at me, demanding to know why I didn't kick, scratch or bite the attacker's ear off? I could have said, 'Is that what you did when you were attacked? But instead, in shock to his attitude, I said nothing at all. His words haunted me for years. I just had to know why. I imagine you may want to know what answer I discovered, which leads to our next question.

Question: Why didn't I scream and fight back?

Answer:

"…*I am so troubled that I cannot speak.*" *Psalms 77:4b*

"*…because fear has torment…*" *I John 4:18*

I didn't scream because I was terrified. Some time ago I saw a commercial for a home security system. In it a woman was feeding her baby when she noticed a man standing in her yard outside in the dark, watching her suspiciously. She was shocked speechless. The narrator's voice said: "It has been shown that absolute fear can render one silent." I remember feeling relieved when I saw that because I thought for years that there was something wrong with me.

Actually, I did try to fight my attacker at first, but he was stronger and returned every attempt I made by being more aggressive. He overpowered me both physically and mentally. In my rationale, I thought giving up might ease his grip on me. Who knows, my submission may have helped to save me from more abuse. I was not a coward. The instinct to survive made me act with reason. I didn't want to die violently.

Question: I was attacked more than once by the same person, does that mean I was asking for more?

Answer:

" *The thief cometh not, but for to steal, and to kill, and to destroy...* "
John 10:10

If the same attacker comes after you more than once, it is not because you asked for it, but because the enemy's goal is to steal, kill and destroy. His return only acknowledges the fact that there is more to you than he could ever take away. He comes because he could not defeat you. God didn't let him, and He won't. Know that this goes much deeper than how it appears.

In Dinah's case, the man who raped her didn't just want her for the moment, he wanted to possess her permanently. He was not satisfied with a one-time experience with her. Because she was still a child, he talked nicely to her afterwards to convince her. The truth is, he was being used by the devil in an attempt to steal from Dinah's family the wealth and status that God had blessed them with. If that plan had been successful, Jacob's family would have been destroyed because they were already outnumbered.

The enemy is insatiable in his appetite for power. He doesn't just want to steal, he wants to keep coming until he kills and destroys everything he possibly can. He knows that once he does this, it makes God appear that He doesn't care for us, which is a lie.

So when the enemy comes, in whatever way he comes, no matter how many times he comes, he comes to grieve the Lord, steal your security and quality of life, kill your hope and destroy your sense of peace. But he cannot succeed as long as you decide to call on the Lord's name, regardless.

Question: Why did I keep the experience to myself?

Answer:
"...I am so troubled that I cannot speak..." *Psalm 77:4*

"And I, whither shall I cause my shame to go...?"
2 Samuel 13:13

Honestly, I resolved that I could handle it myself. Besides, I was too ashamed to explain and too afraid to expose myself - again. I didn't want to be doubted or blamed. I did not want anyone to add insult to my injury by labeling me, or worse, disbelieving me. I was too hurt to risk more hurt.

Of course, after I did open up, there were those who insisted I keep it to myself. I realized later that the best thing is to tell the right one, that is, the One who can help you.

What I didn't realize was that an invisible enemy had attacked me. Even if I had told everyone I knew about what happened, I would not have been healed. I just would have been regarded differently. I would be 'the one who was assaulted' in the eyes of those who knew. Victim would be my new title, as far as they understood.

I needed the Lord to help me. I had to cry out to Him and before I did, He was already listening. Telling Him would not only bring about true comfort, it would spare me more shame. The Lord would never view me as a pitiful soul, for which to feel sorry. He would love me and restore me to who I am destined by Him to be. No one else could have done that.

Question: I am constantly looking over my shoulder, sometimes checking every closet and room when I get home and rushing everywhere when I am out. At times I have nightmares when I sleep and flashbacks when I am awake. I can't even enjoy certain movies without having to leave the room. When I am alone in a room with a man I don't know well, I am terrified. How do I deal with this fear that I now live with?

Answer:

"I, even I, am he that comforteth you: who art thou, that thou shouldest be afraid of a man that shall die, and of the son of man which shall be made as grass..." *Isaiah 51:12*

"Fear thou not; for I am with thee: be not dismayed; for I am thy God: I will strengthen thee; yea, I will help thee; yea, I will uphold thee with the right hand of my righteousness." *Isaiah 41:10*

"For God hath not given us the spirit of fear; but of power, and of love, and of a sound mind." *2 Timothy 1:7*

Do you believe that God can restore you? If you ask Him to remove the fear, He will. Then He will give you a love for life itself. So ask Him to give you the power you need to overcome negative and hindering thoughts. Pray that He gives you a sound mind so that you won't be jittery and jumpy. It's okay to ask the Lord for help. He wants to help.

Now take time to consider the scriptures listed previously. Once you learn them, you will be ready to confront and defeat that spirit of fear with the Word of God, by the Spirit of God and in the name of Jesus Christ, the Son of God. From now on, whenever you feel afraid, think about these scriptures and focus on them instead of the object of your fear. This is how fear will be dispelled so that you can go on with your life in peace.

"Submit yourselves therefore to God. Resist the devil, and he will flee from you." *James 4:7*

Resist fear by doing the opposite of what fear tells you to do. Go out when fear tells you to stay in. Walk when fear is pushing you to run. Stay when it tells you to go. When fear can no longer push you around and dictate your life, you will be free from it. Oppression cannot operate without fear. It has to have a means to hold you and that is the torture chamber of paranoia. Thank God you don't have to live that way. You have been oppressed long enough. Pray this prayer found in the Bible:

"Hear my voice, O God, in my prayer: preserve my life from fear of the enemy. Hide me from the secret counsel of the wicked; from the insurrection of the workers of iniquity." *Psalm 64:1-2*

Question: I cannot get past the defilement of my body and my soul. I feel like I have lost full rights to my own body. I have been reduced to a mere shadow of who I used to be. How do I get past the shame and disgrace that I feel?

Answer:

"Draw nigh unto my soul, and redeem it: deliver me because of mine enemies. Thou hast known my reproach, and my shame, and my dishonour: mine adversaries are all before thee. Reproach hath broken my heart; and I am full of heaviness: and I looked for some to take pity, but there was none; and for comforters, but I found none." *Psalm 69:18-20*

This passage of scripture is so very beautiful. I am so moved at how much God cares about us. Our hearts are important to Him. He doesn't just demand us to 'Get over it!' He feels for us. This passage is actually describing the thoughts of Jesus Christ Himself as He voluntarily suffered the punishment of sinful man, so that He could save them. See, the Lord truly does know how you feel, He has been there.

You must realize that what you feel is only due to the broken state of your heart. No one except the Lord can help you completely. Cry out to Him. He is listening. He wants to help you and remove every ounce of shame, guilt and dishonor. The Lord wants to bring you to a peaceful place where no one will trouble you again, like He did with Dinah. God is there to save you and make you over again – inside. Just mention His name. The Lord is ready to rescue you, you cannot rescue yourself…no one can.

For now, this brings our silent conversation to an end. One day, perhaps, we will meet, you and I, but until then, just know that I am praying for you. I know that I cannot relate to you in all aspects, but I do know that the Christ can. He is the Savior. You can talk to Him whenever you want to and He will hear you when you speak to Him in faith. He will even answer you.

"Call unto me, and I will answer thee, and shew thee great and mighty things, which thou knowest not." *Jeremiah 33:3*

What a great comfort to know that God will answer us when we call Him. He is the answer to our problems. So take heart and have peace- your hope is in the Lord.

" I waited patiently for the Lord; and he inclined unto me, and heard my cry." *Psalm 40:1*

Chapter Three: The Chastisement of Our Peace

I think I know how you must feel. When no one else heard your scream but God, because your scream was inaudible. Or when the whole world heard you - - the birds heard you and so did the sky, but no one responded. Or even…even the cry that was heard too late. I think I know how you must feel.

When your senses learned what perversion is. When a part of you or even all of you was touched by one that had no right to. When your palate was soured by a most awful taste and your nostrils inhaled a stench most evil. Your eyes beheld an embodied spirit of destruction and your ears could actually detect the sound of every one of your heartbeats clear as a drum. When you learned what it meant to be overpowered by someone else, bearing the pinning weight of someone introducing you to oppression. I think I know how you must feel.

When you discover that soap and water can't wash away all types of filth. When harmless neighbors become potential enemies. When the lights of your soul are turned off and you are wide awake. And what of those dreams that come day or night even when you're not completely asleep? Or the sudden vulnerability to anyone who seems even slightly more powerful in stature? I think I know how you must feel.

When no place on this earth is a guaranteed haven. No one seems completely trustworthy anymore. Everyone seems to be either helpless to rescue you or is out to harm you. Every direct stare or those who stand too near is a threat. Even silence can be haunting. I think I know how you must feel.

When you wonder if a day will come when you won't remember. Or to be unmoved when you are "reminded". When you want to be a child again – just innocent and free – not looking over your shoulder. Not avoiding anyone at all. To be able to be what you were before you were changed – and yes, you have been changed. I think I know how you must feel.

To read about yourself in the paper…and not even have a name. To hear your neighbors and colleagues talk about "that poor victim", without even realizing that you are the victim they are speaking of. To stand before strangers who accuse you in their minds of causing your own pain because you lacked, I guess, "super" power. To have to tell your story ten or a million times and never being able to clearly express it even once. Or to never tell the story at all and experience no less pain. I think I know how you must feel.

To wonder if your eyes will ever feel normal or will they always be sore from secret crying? To wonder when the bruises will go away. And when those who know, will ever stop looking at you like that, or better still, cease from avoiding direct eye contact with you at all. I think I know how you must feel.

All of that said, although I think I know how you must feel, I can't do anything to ease how you feel any more than I could help the way I used to feel- - except to testify to you that I know of someone who can.

It was Jesus Christ who did what no counselor, mentor, psychiatrist, therapist, parent or friend would have ever been able to do. The talk shows had a lot of talk – but not one clear word that really helped me. The magazines and books could re- count and suggest but none could heal my broken heart. The professionals listen with great concern, but none could turn my life around completely and give me a new start. Wisdom is to be prized and knowledge is great, but healing is what was needed more than these.

Why did it happen? Why was I ravaged relentlessly? Because the world is full of cruelty and evil. Until Jesus returns, this is the way the world is for those who don't know about the power of God for their lives. Why me, a Christian who loved the Lord and was even then filled with the Spirit of God and serving in ministry? Because, at that time, I still hadn't learned that through Jesus Christ, I am victorious over the evils this crucifying world can bring. I do know better now, and I have no regret. See, what God has caused to take place in me since that experience, may help millions of others in a way only one "who has been there;" perhaps, can deliver.

God truly worked all of it out for the good. You, I prayerfully believe, will be one of the recipients of that good. I don't believe you would be reading this book unless you were or someone you know was harmed sexually. My hope is that you will soon have the same newness of life that I have been given in Jesus the Christ and you will be a better person than you ever imagined. Better for your own good and better for others.

There is a certain kind of peace we live with that stems from the security we find in our homes. It began when we were first swaddled in the arms of our parents and family, then continued as we were surrounded by those who stood ready to catch us, as we took our first steps. It strengthened as we learned that we were protected, sheltered and heard. We knew that we were valued. We lived with the peace that comes from being loved.

I believe that Dinah had this kind of peace during her entire childhood. She was the only girl amongst a house full of males. Her name means vindication, so you can imagine how cherished and secure she must have felt. The day she walked through the new land to find other girls was no different to her than any other day. She had known kindness all of her life and she no doubt expected that from her new neighbors. She was at peace. You and I were born with the same innocent, trusting- - peace. See, the attacks that happened to me years ago were to undermine and nullify my peace.

The enemy of our souls will try any method he can to steal our peace. He is aware that such serenity is a pathway to the Prince of Peace, who is Jesus the Lord. It is our nature as humans to gravitate towards those whom we are like. When we are very young, we have great peace. Because of this, it is very easy for us to believe in the Lord, for He is our peace. We are drawn to the Lord as soon as we learn about Him because His peace is what we have in our young innocent hearts. We identify easily with this Prince of Peace, unless, through some misfortune, we lose our peace.

That's where the enemy of our souls tries to come in. The devil wants to disturb our peace so that we will not want to trust the Lord. Though sickness, poverty and hatred are some of the tools the enemy often uses, sexual crime, second only to death, is one of the strongest weapons he wields to combat us.

This is often why he attacks children, while they are still naïve to the woes of the world. He takes advantage of their trust, to crush, if it were possible, their propensity to seek after God.

Virgins, like Dinah, are another prime target of the enemy. To ruin a virgin is to possibly end a bloodline, a race, a nation and even more importantly, a deliverer from coming forth into the earth. A deliverer who, through the power of God, is capable of leading thousands away from the bondage of the devil and into the loving arms of the Lord. I reiterate- - it really isn't personal.

Through sexual crime, the devil can bring fear, grief, sorrow, hurt, confusion, anger, hatred and doubt all with one blow. All while eliminating the natural sense of peace that we knew all our lives. This is why so many that have been affected by sexual crime mourn so severely. Much was lost. To be overpowered by another human into a state of momentary slavery does something to the soul that is inexplicable. It is inconceivable to others to even begin to discern the level of damage such a horror as rape can produce.

But we can thank the Lord that through the promises He gave to us in His written Word, we can recover all that we have lost, even our peace. Peace is often lost when something bad happens to us.

At other times, we tend to keep peace at arms distance whenever we allow a thought of what 'might happen' run rampant in our minds.

We cannot let that happen. Our minds are where a lot of our biggest emotional battles are won or lost. We have to come to grips with what happened and not let it redefine us.

"There is no temptation taken you, but such as is common to man: but God is faithful, who will not suffer you to be tempted above that ye are able; but will with the temptation also make a way to escape, that ye may be able to bear it." *I Corinthians 10:13*

There is a factual resolve I hold. Spiritually speaking, I am not going through any more or any less than any other person on the face of this earth. Because strength of spirit varies, so does the impact that tragedy has on individuals. One tragic event may bother one slightly, but the same type of experience may devastate another person entirely. Therefore, no one can really say, "I know how you feel;" they can only imagine and empathize.

God alone knows what will shake each soul to a point of no return. He knows that the sin/curse that was released in the world because of Adam's defiance, affects and will continually affect everyone. However, through the Lord's mercy, He sovereignly keeps the devil from tormenting everyone to a dangerous point of madness. Only to the point of decision are we driven.

There are several passages in the bible that show us how important peace is. In one passage it explains the sacrifice that Jesus made so that we might have real life with Him. In order for us to have peace,

Jesus endured cruel punishment to correct the problem of our unrest. This is what it says:

"...The chastisement of our peace was upon Him..." Isaiah 53:5

This scripture is very comforting. When the Lord gave His son Jesus to die for the penalty of our sins, He also allowed Him to pay for our earthly suffering as well. Jesus knew that we would encounter all manner of evil that would rob us of peace. So He made provision for us. He took upon Himself the chastisement (or punishment) that we would have been suffering with our whole lives. In doing so, Jesus made the way for us to have peace instead. This way, we are never completely devoid of peace. It is always, thank God, available to us.

Whenever we do experience something that takes our peace away, it is a tool the enemy uses to knock us down. However, Jesus took the brunt of that blow so that it would not keep us down. If we trust Him, we will have peace again. It doesn't matter what we are going through, when Jesus is our Lord, we will have peace even in the midst of our trouble.

Not just the peace we have when everything is okay, the peace that keeps our mind at ease and our hearts calm even when things are not okay. Jesus is the Prince of Peace. He rules over it. He is the one in charge of those good moments when all we can do is sigh in true comfort and ease. When we witness beauty, innocence and calm all at the same time. When there is nothing on our minds but goodness. That is what peace is and that is why we must pursue it.

We must not permit ourselves to live without peace. Only those who choose to live wickedly are they who will not have true peace given to them. Those who follow the Lord will know real peace because He is our peace. God will cause us to sleep at night and really rest. He will remind us that He is with us and we need not be afraid. He will reassure us that we are protected and provided for. The Lord keeps us safe when we truly trust Him. I can go on and on, but the following passages say it much better than I can explain.

"My presence shall go with thee, and I will give thee rest." *Exodus 33:14*

"I will both lay me down in peace, and sleep: for thou, Lord, only makest me dwell in safety." *Psalm 4:8*

"Peace I leave with you, <u>my peace</u> I give unto you: <u>not as the world giveth</u>, give I unto you: Let not your heart be troubled, neither let it be afraid" *John 14:27*

"These things I have spoken unto you, that in me ye might have peace. <u>In the world ye shall have tribulation: but be of good cheer; I have overcome the world."</u> *John 16:33*

See, these verses are only a few of the words that God spoke to those He loves. You are one of them. So be at peace. Jesus already paid the price for what sexual crime has done to you. He purchased back your peace years before you even lost it. Accept His credit (so to speak), by believing in Him. Trust Him, He will not disappoint you.

In Ephesians 2:14 it says *"...for He is our peace."* This is what we must focus on the next time we remember something horrible from the past, feel threatened in the present or bleak about the future. We are not alone. When we acknowledge God and keep our thoughts on Him, the enemy <u>cannot</u> take our peace away.

So keep praying. Keep seeking after the helping hand of God. Keep asking God to heal you every time the very thought of what happened occurs to you. You will see your life coming to a place of newness more and more each time that you do.

"For the weapons of our warfare are not carnal, but mighty through God to the pulling down of strong holds; Casting down imaginations, and every high thing that exalteth itself against the knowledge of God, and bringing into captivity every thought to the obedience of Christ..." *2 Corinthians 10:4-5*

As you pray to the Lord to restore your life, consider what part you can do to rebuild yourself. I strongly suggest that you find expressive avenues of release. Dance, art, music, poetry, gardening, rock-climbing, writing, sewing, exercise, volunteering, cooking or whatever creative thing you enjoy. Do what you like or would like to try to do. A new craft or skill is a small way of releasing yourself from internal isolation and stress.

There are only two directions to choose from as you walk away from that tragic experience and on to the rest of your life. The way to life and peace is one direction. The other direction leads to the exact opposite. Look at the picture that follows this paragraph. This is where many victims are left: in a state of hurt and indifference that God never intended.

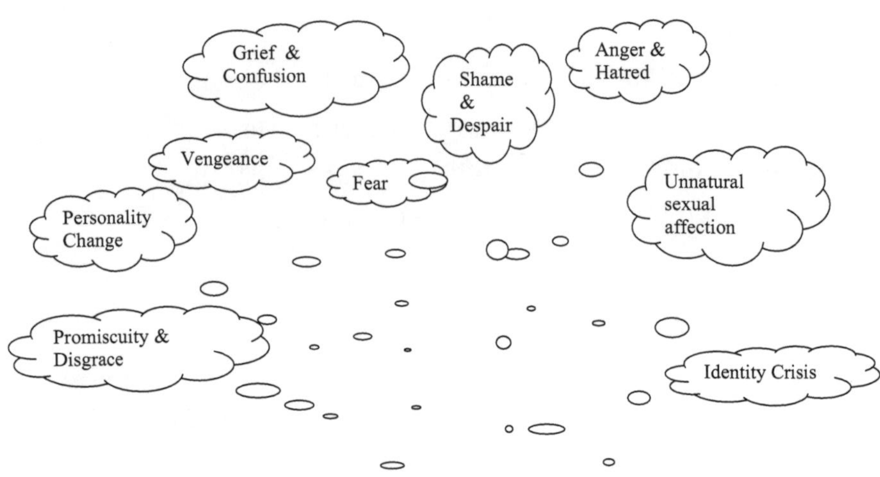

VICTIMS

God wants you to move away from the common reaction of victims, hiding your pain. Return to who you really are by embracing new life and peace in Christ Jesus.

God is so good, He refused to allow the enemy of our souls to keep us from our dreams. He would not let anything stand in our way. If we are yet alive, we have hope, if we choose to accept it. Take comfort in knowing that your peace is just a prayer away. Like He did for Dinah, our heavenly Father can make it so that the predator can never come near you again.

Journal Entry About God's Comfort:

Family & Friends

In this first section, we have examined in detail the life of Dinah and the powerful truths her account holds for us.

Because her family's actions were told so vividly, we all can relate to the range of emotion that they endured.

What we gain from this account is that God has every intention to intervene in the enemy's unquenchable thirst for destruction.

Knowing this, we have to allow ourselves enough sensibility to let God avenge. Our efforts at vengeance can only cause more problems.

Follow the right path. Help your loved one get through this by showing them scriptures of comfort, praying for them and when they are ready—with them. Listen to them. Talk about things that they love to do.

Everything may have changed since the attack, but this change is only temporary. A new zest for life and appreciation will emerge if the right battles are fought at the right time.

Be sure the time of true consolation is not neglected. Aside from the justice, real peace and quietness is a strong need. Don't let your dear ones seek it from someone or 'something' else.

section two:

"...to preach the Lord's ***favor*** *and the day of* ***vengeance*** *of our God..."*

"And it came to pass, after the year was expired, at the time when kings go forth to battle, that David sent Joab and his servants with him, and all Israel; and they destroyed the children of Ammon, and besieged Rabbah. But David tarried still at Jerusalem.

And it came to pass in an eveningtide, that David arose from off his bed, and walked upon the roof of the king's house: and from the roof he saw a woman washing herself; and the woman was very beautiful to look upon. And David sent and enquired after the woman. And one said, 'Is not this Bathsheba, the daughter of Eliam, the wife of Uriah the Hittite?'

And David sent messengers, and took her, and she came in unto him, and he lay with her; for she was purified from her uncleanness: and she returned unto her house.

And the woman conceived, and sent and told David, and said, I am with child. And David sent to Joab, saying, 'Send me Uriah the Hittite.' And Joab sent Uriah to David. And when Uriah was come unto him, David demanded of him how Joab did and how the people did, and how the war prospered.

And David said to Uriah, Go down to thy house, and wash thy feet. And Uriah departed out of the king's house, and there followed him a mess of meat from the king. But Uriah slept at the door of the king's house with all the servants of his lord, and went not down to his house.

And when they had told David, saying, Uriah went not down unto his house, David said unto Uriah, 'Camest thou not from thy journey? Why then didst thou not go down unto thine house?'

And Uriah said unto David, 'The ark, and Israel, and Judah, abide in tents; and my lord Joab, and the servants of my lord, are encamped in the open fields; shall I then go into mine house, to eat and to drink, and to lie with my wife? As thou livest, and as thy soul liveth, I will not do this thing.'

And David said to Uriah, 'Tarry here to day also, and tomorrow, I will let thee depart.' So Uriah abode in Jerusalem that day, and the morrow. And when David had called him, he did eat and drink before him, and he made him drunk: and at even he went out to lie on his bed with the servants of his lord, but went not down to his house.

And it came to pass in the morning, that David wrote a letter to Joab, and sent it by the hand of Uriah. And he wrote in the letter, saying, 'Set ye Uriah in the forefront of the hottest battle, and retire ye from him, that he may be smitten, and die.'

And it came to pass, when Joab observed the city, that he assigned Uriah unto a place where he knew that valiant men were. And the men of the city went out, and fought with Joab: and there fell some of the people of the servants of David; and Uriah the Hittite died also.

Then Joab sent and told David all the things concerning the war;...

And the messenger said unto David, 'Surely the men prevailed against us, and came out unto us into the field, and we were upon them even unto the entering of the gate. And the shooters shot from off the wall upon thy servants; and some of the king's servants be dead, and thy servant Uriah the Hittite is dead also.'

...and when the wife of Uriah heard that Uriah her husband was dead, she mourned for her husband. And when the mourning was past, David sent and fetched her to his house, and she became his wife, and bare him a son.

But the thing that David had done displeased the Lord. And the Lord sent Nathan unto David. And he came unto him, and said unto him, 'There were two men in one city; the one rich, and the other poor. The rich man had exceeding many flocks and herds: But the poor man had nothing, save one ewe lamb, which he had bought and nourished up: and it grew up together with him, and with his children; it did eat of his own meat, and drank of his own cup, and lay in his bosom, and was unto him as a daughter.

And there came a traveler unto the rich man, and he spared to take of his own flock and of his own herd, to dress for the wayfaring man that was come unto him; but took the poor man's lamb, and dressed it for the man that was come to him.'

And David's anger was greatly kindled against the man; and he said to Nathan, 'As the Lord liveth, the man that hath done this thing shall surely die: And he shall restore the lamb fourfold, because he did this thing, and because he had no pity.'

And Nathan said to David, 'Thou art the man'."

2 Samuel 11:1-18, 22-24, 26-27; 12:1-7a

Chapter Four: Defenseless

We begin this chapter by elaborating on the scripture that is recorded at the beginning of this section. What followed this passage of scripture was the record of all the calamity that arose in David's life immediately afterwards. God dealt severely with David, but let him live. David repented, and God forgave him, even though the rest of his life was very chaotic, even until the very day of his death.

Because of his repented heart, many have heralded David as a sort of hero because he was wise enough to acknowledge his sin and get right with God. They relate to him because of the way he humbled himself and pleaded for the Lord not to leave him (See Psalm 51). To see God's response to David encourages many of us because we all need to pray that type of prayer at some time.

David, however, is not who we want to look at right now. What about Bathsheba? What about her husband Uriah? What is to be said in defense of their case? Bathsheba has for years been accused of seducing King David and causing him to fall. What was her defense? Or is she defenseless?

Uriah can be falsely described as a workaholic who values his job more than his family, so much so that his wife was deprived of affection and was left to seek attention from another man. This is just not true, but what is his defense? Or is he defenseless as well?

A defenseless person is someone who has no voice. Someone that cannot protect himself, who has no one to guard him or watch out for him. He is vulnerable and open to attack and harm. He can't speak on his own behalf and plead his cause. He has no argument. He is not strong enough to fight for himself. He possesses no weapon powerful enough to stop an assault. He is without the necessary tools to keep himself untouched. He is weak only by means of the fierce strength of his enemy.

In society, those who are defenseless and weak are usually children, the elderly and women. Men are naturally stronger than women – that is not contested. Adults are stronger than young children. The younger adults are generally stronger than the elderly. Of course, this is only pertaining to the physical dynamic. Anyone who purposely harms one who is physically weaker is the one who is really weak.

It is the job of the defense attorney to plead the case of those who are unable to defend themselves in court. He presents the case to the judge and jury in a way that will prove the innocence of the one he is defending.

Many times I have heard the story of David and Bathsheba told from the standpoint of the woman being guilty of exhibitionism (exposing herself) and thus causing all of the calamity that followed to occur. How often have I heard the question; "Why was she bathing where the king could see her?" "She must have been trying to ensnare him." "It was her fault for being so loose." "David was just fine until she ruined his life." "Women like Bathsheba are the downfall of many men."

I could go on and on with more accusatory statements that I have heard about Bathsheba, but I won't. Bathsheba is not here to defend herself. She cannot speak to her accusers to explain her actions. The truth is, she does have a defense. The very story itself as it is recorded in the Bible is full of evidence of what really happened. It requires us to study a little and look beyond the surface of our first impression of this story.

As the Lord directed me to follow this story closer, I learned that very little of what I assumed was what actually occurred. The wording is very specific and detailed. However, the meanings of the words are sometimes lost in the various translation processes.

So when I began to follow the Lord's prompting to really look into this story, I was amazed at what I discovered. I simply looked up words that I thought I knew the meaning of, in the Hebrew language concordance, because this passage was originally written in Hebrew. With all the various meanings of several of the words, I was astounded with what I found.

The story of David and Bathsheba, is much more than the supposed account of a man who fell into the seductive wiles of a loose woman. Neither is it a case of adultery, as this offense requires two consenting adults that have a relationship with each other. It is a case of voyeurism, authoritative rape, conspiracy, witchcraft, pre-meditated murder, abuse of power and deceit. The very words in this scripture are all the proof we need to answer in Bathsheba's defense.

In case you are wondering why I even want to defend the honor of a woman who lived thousands of years ago, just bear with me. Bathsheba's story is really no different than that of countless others who have only God as a witness to what they endured. They may be defenseless right now, but the Lord will avenge. I have a special interest in those women who, like Bathsheba, ended up marrying the very men who raped them, even though in their hearts, they did not want to.

Just as the Lord moved me to study this woman's case, He will likewise see to it that the stories of so many others are justified as well. Perhaps this story told with regard to Bathsheba and not David, will do just that. God allowed all the details of this account to be recorded for a *reason*. The details are what matter in matters of justice.

On behalf of Bathsheba, I would like to submit my findings as evidence to support her case. After all, she represents women and men from all over the world that are without defense. No one knows the real story of these countless victims except the Lord, and He does not ignore injustice.

I am not an attorney, but I can present this evidence of Bathsheba and David's case, so to speak, to a jury of readers who may, like Bathsheba, be without a voice.

We begin this inquiry by directly examining what the scripture reads from beginning to end. Look again at the text and pay close attention to the words that are highlighted. These passages and the words themselves are what I want to submit as evidence.

The first word we will look at is <u>bed</u>. At first I thought it was what any one would think, a place to sleep. But when the Lord prompted me to study the original translation, I followed through. What I learned is that there are actually two distinct words for <u>bed</u> in the Hebrew and Chaldee languages. One is a bed that you sleep in while the other is a bed for carnal, sexual purposes which is the word 'mishkab'. It means carnal intercourse bed, couch. This is the type of bed that David arose off of (when he should have been at war with all the other kings). So most likely his frame of mind was on sexual meditations.

Now keep in mind, David was a married man at the time. Should a carnal intercourse bed have even been in his house? Furthermore, he was a man of God and the King of Israel. David should have shunned this type of excess simply because he should not have conducted himself like a heathen king.

According to other references in the Bible, we learn of the behavior of the carnal kings of that time. In the book of Esther, the King got rid of his Queen when she displeased him and acquired a new Queen. Both Abraham and Isaac had very real fears that the king would take their wives, because of their beauty, and they would be killed. The kings in that time in history could even have as many wives and concubines as they so desired. The sexuality of a king appears to have been a celebrated thing in those days.

But a king that was a servant of God should have had a different standard. He was to abide by what was right and just; using his power to indulge in whatever lustful cravings his body demanded was neither. Evidently, the haughty temptation to be like the other kings was perhaps too great for David.

Like other kings in the land, David had several wives and concubines. Apparently he had fallen into the idea of celebrating his sexuality. This is clearly seen by how easily his men obeyed his wishes to forcefully take a woman from her home to satisfy his sexual edacity. And like a heathen king, he had no problem killing Bathsheba's husband to make her his own wife.

The very fact that David had in his possession a 'mishkab' or (carnal intercourse bed or couch) gives concern regarding his sexual discipline. But he did not stay in it, he got up off of the bed and walked ('halak') on the top of the roof. This word can also be translated "walk to and fro, up and down." So you see, David wasn't just taking a stroll for a breath of fresh air. He was pacing back and forth, refusing to leave that area of his house. We can clearly understand why when we look at what the word saw means.

David saw the woman bathing, and at first, I assumed he was just glancing over at her or that he just so happened to notice her. In all actuality, David did a little more than that. The Hebrew word that was used in this text to describe how he *saw* her is 'raah', which means to see, gaze, spy or stare. It's not the word see or even stare that concern me, but it is the word spy that gives great alarm. It is indicative of intention, that is, he deliberately made an effort to invade the privacy of this woman. I have never known or heard of anyone who spied without the purpose of acting on his findings.

To *gaze* is also a strong term because it goes beyond just staring. See, you can stare at something and actually be thinking of something else in a daydream like state of mind. But gazing is when you fix your eyes on something earnestly and intently. David wasn't even trying to look away. The thought of his wives undoubtedly didn't cross his mind. He was fixated on this woman.

One would think that this woman must have been a really good-looking person. After all, the passage does say that Bathsheba was "*beautiful to look upon*". However, this word happens to be a key part of the evidence that she was not as many believe.

The Hebrew word for *beautiful* in the text is the only time that I found it used in the entire Bible. There are other Hebrew words used that are translated to mean beautiful, but only one in particular that was used to describe Bathsheba. It is the word 'towb' and it is not a word that has very much to do with physical appearance.

The definition of the word "towb" in the Hebrew language is cheerful, at ease, fine, loving, merry, joyful, kindly, pleasant, precious, approachable and sweet. It denotes nothing about her sensuality at all. It is a word that primarily describes inward qualities; not a word that has to do with looks alone. Besides, David had many women at his disposal. A bathing female was no doubt nothing new to him. There had to be something different and unique about her far above physical appearance that attracted this great king.

A rare woman Bathsheba was, when you think about it. She was kind, sweet, pleasant and cheerful. She was loving, at ease and, get this, she was approachable. Who wouldn't want to be around someone like that? Is it possible, that David had never met anyone like that before? Did the very idea so impress this man, who, by the way, had many wives, that he felt he must have her?

So when we consider Bathsheba in the light of the word that is used to describe her, we really have to say that it was quite possibly her demeanor that got David's attention, not merely her physique. A woman who has inward beauty like that is indeed one to take notice of, but not to take advantage of.

We know that David's wife Michal disrespected him. David did not woo her; He earned her as a war trophy agreement made by her father. Another wife, Abigail, whose hand he did ask for properly and kindly, was one who did respect him. However, he may have felt obligated to marry her. Her husband died after he found out that she secretly took it upon herself to honor a request made by David that her husband foolishly rejected. At that time, David was not yet king, but he was already anointed by God to be the future king, compelling Abigail, out of respect for God, to help him. So David married Abigail when he learned that her husband had died suddenly. Perhaps he wanted to make sure that he provided for this woman who had provided for him. Then, there was another woman he married named Maachah who was the daughter of King Talmai of Geshur. Their marriage may have possibly been a political agreement.

Now, the one named Eglah was the only one I found listed in his family record as his "wife" even though the other women were named.

It is thought that she was perhaps his original wife. Then of course there was Haggith and Abital as well. Ahinoam was the wife who gave him his firstborn son. In fact all of his wives, with the exception of Michal gave him a son.

I found no record of the details of his many concubines, but we can imagine that the physical aspect was quite possibly the main purpose for David to secure them as his property.

This brings us to the state David's mind must have been in when he saw Bathsheba. She was not disrespectful and vain like his wife Michal, she was approachable and pleasant. She was not someone that needed him, as his wife Abigail did, but was joyful and fine. She was not one who just met a physical desire, like his nameless, possibly reluctant, concubines, but was loving, sweet and precious.

David grew up with a lot of rejection. His own father didn't even consider him to be worthy of being anointed as king. His brothers didn't take him seriously even when he spoke words that glorified God. His predecessor, King Saul, tried several times to kill him because of jealousy. He even gave his daughter, David's espoused wife to someone else. Of course, David used his power later to get her back. She came to despise David and he cursed her to a childless life because she dishonored God by berating David for how he worshipped the Lord. (See 1Samuel 25:44 & 2Samuel 3:12-16; 6:16-23)

It is very possible that a woman as kindly and sweet as Bathsheba appeared to be, was the type of woman that David wanted all his life. When he learned that she was married to one of his own valiant and decorated soldiers, he decided to use his authority to just… experience her. He wanted to just touch what he knew he couldn't have—even if for one brief moment.

It was an issue of power. For so long he had been denied the love and respect that any one desires. After all, what did he ever do to anyone? He had helped so many and was mightily used by God to deliver His people time and time again. He had always behaved himself wisely and done what was right.

Now that he was finally king, after years of struggle, I suppose he felt that he should at least get some of the fringe benefits. If he wanted a carnal intercourse bed in his house like the other kings, why not? If he wanted more concubines than he knew what to do with, so what! If he desires to stay home and not go to work, why not? He was the king after all! No more denial, no more disrespect. He had a good bunch of men working for him. Men like Joab, who would do anything for him. He had wealth. He had the admiration of the people, why not have it all? Who can say no to the greatest king in the world?!

Who knows, this may be the kind of thinking that David pondered when he paced back and forth on the roof he should have never been on in the first place. When he saw Bathsheba, she no doubt reminded him of all he had not conquered. A woman of his choosing. Settled. Joyful. Approachable and therefore, respectful. Established and even known by virtue of the high-caliber type of man she married. A real *woman.*

She was the best kind of a woman. A real lady, fit for a king. Not a war-prize who would despise him. Not a widow he felt obligated to support. Not a concubine that he may not have had any connection with because she was a mere possession. Bathsheba was admirable and certainly a sight for the sore eyes of a possibly very disillusioned and insecure man.

So what could he do? Praying for miracles for his own marriages would have been wise, but that wasn't what he did. He decided to appeal to his ego instead. He was king and he could have whatever and whoever he wanted. He could even have her. He saw a lot that day. He saw the desire of his soul personified and he wasn't going to just let her be, he was king, afterall. He could do whatever he pleased. Who would dare say no to him now? Who would cross him? Who would be foolish enough to refuse him?

David did not woo Bathsheba...he took her. He used his position to order her to be brought to him. If he were not the king, he would never have had the authority to take her in that way. That word *took* is the Hebrew word 'laqach' and it means to seize, carry away, fetch, take away. Any time soldiers come to your home to take you at the King's command, you really don't have much choice. Bathsheba did not just go, they seized her. There was no exchange of fondness or flirtatious behavior. How do we know that she even realized that someone on a roof could even see into her home? Bathsheba was not a willing participant and the next key word proves this further.

Once she was in the king's home, the Bible says that "*he lay with her.*" This word lay is translated from the Hebrew word 'shakab', which means to ravish, to cast down (or throw with force), to make to lie down, to make to lie still; to lie with for sexual connection.

Why, if Bathsheba was (supposedly) the one who was seducing him, did he have to force her to lie still? Why did he have to send his soldiers to carry her away to come to him? Why did he have to throw her down with force? This was not Bathsheba's doing. This was all David's action and it seems very obvious, in this light, that he raped her. It isn't too hard to believe that David was capable of the reprehensible crime of rape because he soon after showed that he was capable of pre-meditated murder.

Think about it; she did not welcome him or invite him. She did not send love notes for him to come over and have a sexual tryst. The very words that are recorded discourage that line of thinking. He used his power to ravage her. She seemed to personify everything he may have felt he deserved and had not received. For all his ambitions, efforts and struggles over the years, here was the prize of all prizes, the ultimate conquest: a woman truly worthy of his prestige.

It's interesting that when he was finished, she left and went back home. The word *returned* is used to be exact. If she wanted to be with the king, if in fact that was her goal, why did she leave? Even more curious, the word returned is translated from the word 'shuwb', which means to retreat, rescue or draw back.

Why did she need to be rescued? Why would she draw back if, indeed, to be with the king was what she was aiming for? It did not read that he sent her away, she made the attempt and left. He apparently made no attempt to hold her. And why should he, when she lived right within his reach whenever he had a whim? Had this been mere adultery, a relationship would have been established or at least, an effort to create a continuance of the *affair*, if in fact that is what it was.

Bathsheba was not an adulterer. She was the wife of one of King David's honored men and the daughter of one of his most valiant soldiers. Undoubtedly, she had been a very close neighbor most of her life. Yet before that night, he probably didn't even notice her.

If Bathsheba had wanted the affections of David, being a military daughter, military wife and close neighbor, she would have had many opportunities to do so. Not only that, because it was a season where all kings go to battle, there was really no reason that this woman could have been aware that the king was in town. Knowing that the men were gone, or at least assuming that the men her husband fought with were gone, she may have been a little less concerned about who was in her midst. Perhaps she didn't realize that the king was still around, and may have been a little too relaxed.

Up to now we have examined the case with regard to the unspoken side of this story. We have had a chance to reexamine the words that we too easily assumed we have understood, and have found that the deeper meanings explain the truth more clearly. Let's continue.

To show Bathsheba's character further, we can see that when she heard that her husband Uriah died, she "*mourned*" for him. The translated word is 'caphad' and it means to lament, wail, tear the hair and beat the breasts in grief. This doesn't sound like the action of a woman who did not love her husband. If we look at the parable that the prophet Naaman used to describe her marriage to Uriah, she was like "*a beloved lamb, the only lamb of one who cared very much for her".* She was not just a wife to Uriah, he adored her. That analogy also describes Bathsheba as lying in the bosom of her husband.

I don't think I need to mention the fact that many women are adored by men that they do not love in return. For one to lie in the bosom of another is a sign of adoration and intimacy. Therefore, we can be assured by the Prophet of God's description of her, that Bathsheba and Uriah loved each other. It was a mutual experience. Their love was not one-sided. This was not a man who loved a woman that did not love him back. It was a precious marriage. Uriah treasured and respected her. No wonder the dear woman grieved so when he died.

Uriah did not treat his wife as an object of sexual gratification. When he came to town on the orders of the king, he chose to follow what was moral instead of what was common. He honored the Lord first, even before his wife or his own manly passions. He could not, in good conscience, have a time of intimacy with his wife, while the ark of the Lord was displaced from the sacred place where it belonged. Uriah's marriage was sacred, but not more than his love for God. Uriah was a true worshipper.

How could any real woman not honor a man like that? She didn't even disturb him by sending word to him when she had been harmed. When she learned she was pregnant, she again dealt with the anguish alone. She knew that her dishonorable pregnancy could cause her to be executed. She told King David that she was carrying his child, perhaps hoping he would show mercy and either hide her or grant some level of clemency in order to stay the people from punishing her.

Obviously, she loved her husband too much to trouble his mind while he was on the battlefield. He certainly could not leave his work. Why cloud his judgment with the horrible news of her tragic assault by the king? Uriah couldn't have acted on whatever vengeful feelings he would have had anyway. What could he do to the well-guarded king except show the same subservience as Bathsheba?

It would have been too dangerous for Bathsheba to tell her husband what happened, because it could have cost him his life. That news would only cause him to lose focus on his work and fall into negligence, or break military law by abandoning his field position to return home to be with her. All she could really do was wait for him to return to her arms again and hope the king would be decent enough to protect her (at least for the sake of the unborn child), until then. Bathsheba not only respected her husband, she obviously valued his life, lovingly placing his safety above her own emotional needs.

So what can we say after mulling over a story like that? How is this account relative to today? Why even bother to take so much effort to study all of this in such detail? Two reasons.

One, because thousands upon thousands of victims live their days out suffering secretly with no one else but God and the assailant having knowledge of what really happened. It's hard enough to deal with the assault, but not having anyone to talk to, not being able to release the pain and keeping it all inside just makes it so much more grievous to bear.

How many today keep it all to themselves, never telling anyone what happened? God only knows. To tell Bathsheba's story is to declare that God does, in fact, know about all of us. He let Bathsheba, who was violated privately, to be avenged publicly. Then He recorded the entire account from beginning to end to vindicate her as well. All to show how He can turn even the most dire of situations around for the good.

The second reason that I wanted to highlight this particular story is because many women find themselves married to the very men that assaulted them. They need to know that God sees them too. We will look at this in depth later.

For now, I just wanted to let you know that God does avenge. The record of the story is, in itself, justification for Bathsheba. The same God who showed such acknowledgement to Bathsheba, who, has been historically scorned, will acknowledge and consider us today. God is just that good. As long as God is here with us, we will never be without defense. Sooner or later, truth will always prevail. The Lord will see to that.

"I will lift up mine eyes unto the hills, from whence cometh my help. My help cometh from the Lord, which made heaven and earth. He will not suffer thy foot to be moved: he that keepeth thee will not slumber. Behold, he that keepeth Israel shall neither slumber nor sleep. The Lord is thy keeper: the Lord is thy shade upon thy right hand. The sun shall not smite thee by day, nor the moon by night. The Lord shall preserve thee from all evil; he shall preserve thy soul. The Lord shall preserve thy going out and thy coming in from this time forth, and even for evermore." *Psalm 121*

Chapter Five: Wrong place, Right one

What if Bathsheba did lure King David?? We all, I am sure, have known someone who has demonstrated a seductive or flirtatious lifestyle. The clothes, mannerisms and speech can be very unnerving and even disturbing to observe. Admittedly, it is hard to take seeing someone behave in such a demeaning way.

When you have a loved one who is promiscuous, unwise, partners with questionable friends or has any other risky mannerisms, it is easy to allow thoughts of the worse to enter into your mind. Naturally, you think, that some sick-minded person is going to come along and harm them in some way.

But though the naïve, rebellious or just plain foolish person may frequent the wrong places, because of who you are in Christ, you are the right one to help them. You have the power to speak words of life, protection and victory over them. You can make all the difference in that person's life by being steadfast in praying, believing the word and being sensitive to the spirit as to how you are to intervene, if needed.

Otherwise, you will pace the floor merely hoping for the best but preparing for the worst. Unfortunately, what you believe may come to pass. You will have missed a kinetic warfare moment of prayer.

You saw the worst coming and all you did was shake your head. To cover your dismay, you'll be no doubt tempted to believe that that loved one had it coming. Many may believe that Bathsheba got what she deserved. But no one deserves to be treated that way.

Sometimes, it is common to hear the phrase, "She was at the wrong place at the wrong time," or "He should never have been there." These statements are at times quite factual, but that doesn't mean that the circumstances these statements refer to are the reasons the crimes occurred. Furthermore, it does not conclude the situation.

Why then, do bad things happen to good people? That question can be addressed in many ways I am sure, but in reality, <u>bad things happen to everybody</u>. In Bathsheba's case, I believe the Lord took a bad situation and used it to show just how caring and just He is. He didn't want it to happen to her or anyone else, but because it does happen, He worked it out for her, giving the rest of us hope that He can work it all out for us. She may have been in the wrong place, but being the type of person that she was, she was the right one to be an inspiration for us. She did nothing to have her life so altered, and the Lord certainly vindicated her. Therefore, don't blame yourself for what happened to you. It happened. Let the Lord work it all out for you. You never know how He might want to use you to help someone else.

"Blessed be God, even the Father of our Lord Jesus Christ, the Father of mercies, and the God of all comfort; Who comforteth us in all our tribulation, that we may be able to comfort them which are in any trouble, by the comfort wherewith we ourselves are comforted of

God. For as the sufferings of Christ abound in us, so our consolation also aboundeth by Christ." *2 Corinthians 1:3-5*

Take my experience for example. I spent my adolescence in secret fear. I grew up in a day where the word rape was not common, or at least it was never discussed in front of children. I would note how quiet everyone would get when the news would broadcast an incident of rape. My reaction was just the opposite. My ears would perk up and my undivided attention would be channeled directly into the screen. I wanted to know what I wasn't supposed to know. Why?

My first introduction to the world of passion was really a rude thrust into a world of perversion. There were no rules, just a common goal. The first boys to show interest in me demonstrated it forcefully. They were almost men to me, about age 14 and older. There were at least five of them and one little me; I was nine at the time. That was the last time I was ever automatically polite to strange kids. I thought that bad strangers were big, dirty men with bloodshot eyes. I was wrong, bad has a way of appearing good at times.

I recall that day going on one of my many play-fishing trips at a little water hole in wooded lot near our home. I had my little homemade pole and everything. That's when the boys came from Lord knows where. My older sister Demita who, to my knowledge, had never ventured beyond the playground before, came to the top of the hill at the other end of the lot just before the boys succeeded at ripping my clothes from me. She shouted my name and when she did, her voice startled them just enough to pause, and just enough for me to break free and run like a wild dog was after me. All I could think was that I had done something terribly wrong.

She asked me later what those boys were doing, and I told her they were trying to pick a fight because I knew that she would believe me. (Being targeted by bullies was not a strange thing for me because I spent so much time alone). Anyway, I was just glad she didn't question my story and left it alone, though I wondered if she knew the truth. She never told on me. My very life may have been at stake for all I know, but God had a different plan for me. I thank God that He placed my sister in my life to help me attain it.

Although, by an earnest miracle of God, I escaped that episode in the woods, I could not outrun the spirit of fear. The boys weren't fast enough to catch me, but the evil spirits that influenced them weren't hindered by physical power. For years I grew up terrified that my family would find out and I would be in big trouble. I was more afraid of that than I was of seeing those boys again. I lost my little makeshift fishing pond, clubhouse and safety-pin fishing pole that day. Yet as much as I loved my little corner of the world, I somehow didn't mourn for it.

It seemed like I was a magnet for rude, awkward, pubescent boys for a while after that. An unwelcome touch, smack or squeeze were badges of courage to them. It was almost always boys I did not know, at a pool or school hallway, wherever. I never took it personally, many girls had to endure this treatment, or worse. My few episodes were nothing compared to their daily harassment. I learned how to yell out and slap, and soon found out that angry words rendered less retaliation. I grew to be very cold toward most boys. I saw them as immature- - okay, I thought they were idiots. I had no interest in or respect for many of them. The ones I did like didn't seem to notice me even if we were friends.

I was a pretty normal teenager later on, I guess, but I could not escape that fear. Every time I was outside, day or night, I thought I was being watched by someone who would come out of nowhere, just as those boys had done years ago. I lived with caution whenever I was outside. I learned to sharpen my senses to extremes. I wanted to hear distinctly, pick up a human scent from any direction and see any slight movement from behind me without turning my head. I wanted to remember full descriptions of every person I passed on the street just like I was a police officer. The fear of not being able to describe an assailant was also real. I tried to make it into a game. I pretended to be a martial artist and practiced walking completely silent at my best speed, even in autumn.

That was my way of dealing with this fear as a youth, trying to prepare myself for the worst. I came close before in those woods and I never wanted to feel that helpless again. I had been touched, mishandled and disrespected in a most disturbing way. For this reason my heart goes out to child victims of sexual crimes even more so. Their pain is magnified by their näivete. They don't even know what they have lost and their confusion worsens the pain. They may not speak of it, and too often they blame themselves. Children's experience with sexual crime seems to be so much worse. They pretty much grow up in one moment of time, even though gradual growth has always been the natural way intended by God.

As I grew older and was better able to understand adult terms, I started reading more about the horrible crime called rape. I read every occurrence that I could secretly get my hands on. I found out that victims could be any race, old or young, attractive or plain, male or female. I learned that each experience could be vastly different, yet somehow each was the same. For years I studied the results of reports, research and statistical numbers. I wanted to know the answer. I became, over the years, with this sporadic and inconclusive information, extremely frustrated because none of it was of any help to me. The fear was still there in my mind, only by then, it was worse.

After reading and hearing of so many accounts of sexual crime, I realized that what I had gone through, in comparison to the stories of which I had become aware, was nothing in my eyes. Why then, did I still feel so traumatized? I was wounded just like the scripture says in Proverbs 18:14 that reads, "*A wounded spirit, who can bear?*" I just could not handle it; so I buried it inside of me.

Sure, it would spring up like a bad weed every now and then, but for the most part, I just put it out of my mind. I wish I knew then what I know now about the deliverance of God. My life would have been very different. Even after that wretched day was over and I was left with a wounded soul, I was completely unaware of the fact that God can give me back everything that I lost. It took me decades to learn this truth. I pray that you will come to know and understand this reality before you even finish this book. I don't want you to suffer in silence as I did.

I didn't realize just how much God was willing and able to help me. But just because I didn't know, does not mean that you can't know that truth. I pretty much assumed I was on my own when it came to troubled times. I thought I had to toughen up and just get over it. Try as I may, I could not. Not until I learned that God was there to listen to my cries when no one else would understand, but He is so much more than a good listener. I understood that God was everywhere, I just had no idea that He can really help when we are in trouble. So, regrettably, I lived in a continual state of avoidable wariness.

Eventually, the thing I feared more than anything, that is, the fear that grew in me for years, became reality several years later. I fell prey to that lion I scarcely escaped as a child, and it almost devoured me completely. The Lord saw everything that happened on that dreary Friday afternoon in October. So did I. I wasn't drugged or drunk or in any way unaware of what was going on, I just didn't know how to stop it. Being overpowered and bitten and manhandled was a dizzying and maddening state to be in. Here I was a virgin, my hands held bound by a grown man that was using his other hand to forcibly assault the most sacred parts of my body. That was not how I had imagined my sexual experience to begin.

He stopped and released me when a sound distracted him, only to later trap me to continue his conquest. That was the very next school day, which was after we returned from the Columbus Day weekend. (If I had told someone the first time, he wouldn't have had a second chance, but I was a child… a teenager, but still a child. Telling seemed like an unnecessary thing to me at that time.) That cold, violent, dark moment in my life would have crushed me, but the

Lord stepped in again and halted those dark forces, stopping the rape in progress before he forced himself into me, causing the man to let me go suddenly without a word. My life was in God's hands - -even then. I walked away sore, bruised, mortified and yet--alive.

I can honestly say that the days that followed were almost as bad as the experience itself. The entire situation, the relationship issues, authoritative issues, legal processes and the whole emotional roller coaster was probably the hardest time of my life. Yet, the Lord stayed near me. I recall my rebellion after that; my attempts at promiscuity and the shameful things I did following that assault, trying to prove I wasn't completely powerless. Even in all of that, the Lord still had a way of showing me that He loved me. My boyfriends would later even tell me that God told them not to touch me, (in spite of my behavior). Why was He so concerned about me? Why keep me when so many were left almost in ruin?

Why was I spared after three sexual assaults, still remaining a virgin and later, when I was 20, being stalked by a stranger for weeks, yet not being touched? Or even when I tried to use all of my womanly wiles with men of my choosing, and still God would not let them touch me? Could it be that He really is in control? Of course He is. God is God whether others hurt us or not. Whether we honor Him with our bodies or use them for our own selfish and foolish whims, we still belong to Him.

I wish I had never been touched, unwillingly or *willingly*. Now that I understand that I belong to the Lord, I want to keep what is His-- holy. I have to say this: A rapist cannot make anyone unholy, it is our own sin that does so. When I was under attack during those times in my life, I was still pure in God's eyes. But when I openly offered myself to my boyfriends, trying to get involved with them sexually, that is when I became defiled. My sin is what I was guilty of, even if it did come after the assault, because it was I who decided to behave that way. Now that I am sitting here writing this, I think I know why He did keep me…to perhaps encourage you. Although I was devastated for years, I was still able to go on with my life after being brought down so low and even after bringing myself even lower.

Bathsheba never did act out after David took her. He caused so many problems in her life I cannot even imagine her grief. Why did the Lord bless her and honor her as He did? Because she stayed true to God, in spite of all that happened to her. We should do as Bathsheba did; she is a wonderful example. She continued to be respectful and steadfast. Nothing is ever written in scripture listing her as guilty of anything regarding David. If she could endure the humiliation of being taken from her home, raped, impregnated, widowed and bound in marriage by the same man while still remaining true to God, then we can learn from her today. We can know that God cares for us in every way and just as He did Bathsheba, He will make everything all right.

Please recognize this, even if you were in the wrong place at the wrong time as they say, you are not wrong. As a matter of fact, God would not have even allowed the enemy to come near you if it were inevitable that he would defeat you. You were the right one indeed. So was Bathsheba. So was I. We all have a story to tell that will help someone else escape the clutches of fear and shame. In other words, we endured and are now empowered to help others. How's that for perfect vengeance?

"Blessed be God, even the Father of our Lord Jesus Christ, the Father of mercies, and the God of all comfort; Who comforteth us in all our tribulation, that we may be able to comfort them which are in any trouble, by the comfort wherewith we ourselves are comforted of God. For as the sufferings of Christ abound in us, so our consolation also aboundeth by Christ. And whether we be afflicted, it is for your consolation and salvation, which is effectual in the enduring of the same sufferings which we also suffer: or whether we be comforted, it is for your consolation and salvation. And our hope of you is stedfast, knowing, that as ye are partakers of the sufferings, so shall ye be also of the consolation." *2 Corinthians 1:3-7*

Bathsheba's story is written in black and white to help all that need to learn about the just hand of God. She was not merely a woman who had a bad experience, she was a champion in the Lord. David may have disrupted her life, but the Lord Himself restored it to her with honor. God refused to forget her even after years passed by and David lay dying in his last hours, most likely not giving her another thought. David had suffered for years because of what he did, but Bathsheba, was embraced by the Lord and taken care of with grace until the very end.

If God can do that for her, He can do it for you. He knows what really happened. He knows who did it and why. Rest in knowing that He will avenge perfectly and completely, because you are His and because He loves you.

Chapter Six: "...*or for worse...*"

Of the hundreds of thousands of people who are married today, there are those who have entered into the covenant of marriage for reasons that have had little to do with love. I am referring to the scores of women who married the very ones that sexually defiled them. Because of the shame, this particular group of women may never mention the real circumstances that surrounded them at the beginning of their marriages.

For some, it started out well, with the man being a suitor that she was building a relationship with. All was going well until the day that he forced himself on her. Some women, feeling that they won't be accepted by any other man, marry these. This is understandable in cultures where virginity is insisted upon for a bride. The man may very well have deliberately forced her in order to selfishly reserve her for himself.

Other men will rape only because they have strong imaginations of what an intimate experience would be like with the women they are acquainted with. Their impatience gets the best of them and they refuse to control themselves. Then, once the moment has passed, the man feels remorse and wants to redeem himself by asking the woman to marry him. By this he is attempting to console her and let her know that she is valued.

The woman may indeed follow his plea, believing him to be sincere and marry him. It is especially more likely that she will give in if he blames alcohol or some other alter-influence. He may even try flattering the woman with compliments, thus shifting the blame for his actions on her 'irresistible' beauty.

If the woman is very young or insecure, she may be more likely to marry a man that proposes to her after he assaults her. The woman's parents are key as well. If there is not a strong and open communication already established, the woman may not be able to bring herself to tell them the truth, especially if they can't even discern that she is in distress. They may unwittingly pressure her to go with this man, who no doubt has already won them over with a charming plea for her hand. If poverty on the part of her family is an issue, this pressure may be another weight for the woman to bear if the man is well off.

Many rapists will seek out those they perceive will put forth little resistance. It is therefore not strange to note the odd coupling of mild mannered women with bullish men. Please understand, I am not inferring that marriages where extreme personality differences exist have sexual assault hidden somewhere in the start of the relationship. I only want to bring understanding of how and why some women marry the men that harmed them sexually.

Still, there are other women who fear the anger of the men and believe a worst thing will happen to them or their loved ones if they don't appease them. They marry in terror, never really understanding what type of life they will have by joining a man with such a cruel nature.

That being said, the reasons that women marry the men that forced sex on them can vary from simple ignorance to grave depression and even deep fear. Only the Lord can understand the heart. He alone knows the level of pressure a woman feels when she is presented with such a choice. Her heartache alone may be enough to keep her from being herself and making clear, conscious decisions. After being raped she may see herself as damaged, used or unfit for anyone else. Her dreams of how she wanted her courtship experience to be were shattered and she may give in for fear of rejection by another. This grief is even more intense if the woman was a virgin.

We cannot forget those who do not (or believe that they do not) have any other choice. Culture, age and even conception often play major roles in the way a woman decides. If she feels trapped, cornered or threatened, she has reason to believe that her very survival is at stake. When she is cast out and publicly disgraced, her fate can be sealed indefinitely. What if the rapist is a man of affluence and deceives her family into agreeing to the marriage? For a woman raised in such a culture, how is she to know the way out?

Then of course I cannot neglect to mention the woman who already loves the man so much that the pain he caused in his assault leaves her more confused than anything else. She may rationalize his behavior as a way to forgive the man she loves. She makes excuses for his crime of passion, and may even blame herself for causing him to lose control. She wants to cover him—love does do that.

For all the many reasons women find themselves choosing to marry the men who raped or does rape them, none of them are of any consolation to her once she says "I Do". The only comfort is in the goodness of the Lord.

"I had fainted, unless I had believed to see the goodness of the Lord in the land of the living. Wait on the Lord: be of good courage, and he shall strengthen thine heart: wait, I say, on the Lord."

Psalm 27:13

Is there hope? Of course. There is comfort in knowing that the very God that created the heaven and the earth cares about us. He not only sees our situation, He can actually do something about it. God is familiar with each woman in that type of marriage. He knows the humiliation, the disappointment and the hurt. You must believe this, He will avenge each one and bless them. What type of vengeance will God recompense? Not necessarily as one may imagine.

God's vengeance is perfect and complete. He is not evil nor will He commit evil acts on others. He has established laws, such as the law of reciprocity, the law of sin and death and the appointment of legal intervention. He will punish all who transgress the law because the vengeance of the Lord is righteous.

His vengeance uncovers the secret injustices suffered by the weak. He confronts the offender, exposes his cruel, dark-minded work, and thus vindicates the victim. Another aspect of the Lord's vengeance is His restoration. Regardless of what has taken place, He will orchestrate every detail of our lives and make a brighter future. He will work all things *"...together for good to them that love God, to them who are called according to His purpose."*

Romans 8:28

The Lord will replace what was stolen, locate what was lost and remake what was broken. He will never leave you where you are or as you are. He will change your life. He will never forget you. You matter to Him. Did He forget Bathsheba? No. To the very end God was with her. He changed King David's heart, blessed her with more children and gave her favor with Nathan, the spiritual leader of her time. If that were not enough, He allowed her seed to reign as the next king, saw to it that her story was recorded in detail and mentioned her as the wife of Uriah, the one she truly loved in the family genealogy of Christ. It doesn't matter how or why you married you see, just keep seeking the Lord. God will not ignore what happened to you, He will acknowledge you, consider you and work everything out for you.

To this day Bathsheba is still accused by many of causing King David to fall. For the sake of argument, the question must be addressed-'Why didn't God, who is just, punish Bathsheba?' If she sinned as great an offense as history proclaims, what was her punishment? Why didn't God send a prophet to her with a convicting word of rebuke as He did David? Why did God rename her son a name that denotes God's love? That sure is some favor for a woman who, according to commentators, did such a horrible wrong.

Later on, why did the Lord send the prophet to Bathsheba with specific instructions on how to present her cause to the king as he lay dying, concerning the heritage of her son Solomon? Why not one of the sons of David's other wives, who were older than Solomon and therefore, had legal right to inherit the throne? Why didn't Bathsheba reap from the sin she apparently sowed? Perhaps she wasn't the one who sowed the seed after all.

If this is true, then it explains God's actions concerning her. First, He rebuked David. Second, when she grieved for the child that died as a result of David's wrong, the repentant David started treating her like a true wife and not just a conquered possession. (Comforting her sincerely and getting to know her, not just owning her).

"*And David comforted Bathsheba his wife, and went in unto her, and lay with her: and she bare a son, and he called his name Solomon: and the Lord loved him. And he sent by the hand of Nathan the prophet; and he called his name Jedidiah, because of the Lord.*" *2 Samuel 12:24-25*

Solomon was named by God to take the throne after David's reign. When David was on his deathbed, God ordered the steps of his prophet to go to Bathsheba and encourage her to speak to the king for her son. Even though one of David's other sons was being celebrated as the new self-appointed king, God saw to it that David's final act as king was to crown Bathsheba's son in his stead. Bathsheba's name is recorded in the Old Testament genealogies of King David and even in the Psalms, which is a rarity for a woman. This is an honor and distinction in itself. The same way the Lord honored, vindicated and restored Bathsheba, He can do the same for you. God is no respecter of persons. He is just and He does not change.

In your marriage, you have the hope that this same God who did so much for Bathsheba, who did not have a choice but to marry the king at his command, will be just as favorable to you. Did you know that God's very Spirit is with you to help you walk out your destiny? He hasn't changed.

Bathsheba was raised by a great and valiant man. Her grandfather was a great leader and before David came along, she was married to a man of honor who adored and respected her. She was destined to mother someone great. God saw to it that she fulfilled her calling with her son Solomon. He will likewise see to it that you will realize your dreams as well.

God answered the repentant prayer that David expressed and created "a clean heart in him and a renewed and right spirit." (See Psalm 51) With that, God changed David from being a possessive and violent brute of a man to being a true and considerate husband to Bathsheba. You must believe that God can do the same transforming work in your spouse. All things are possible to those who believe. Just ask Him- -in faith.

God blessed Bathsheba's children, especially Solomon. While the children of David's other wives were involved in all sorts of horrid situations because of his sin, the children he had with Bathsheba were not a part of the confusion. We do know that as a matter of fact, her son Solomon grew up to be a great man blessed with divine wisdom from God himself.

God will bless you as well. He can equip you with wisdom, talent and skill to be all that you should be. He desires you to make something of your life. All you need to do is apply your heart to believe and trust Him. In Christ, we are more than conquerors just because of the fact that He loves us.

If God could do so much for Bathsheba, one who is still accused to this day of seduction, adultery and exhibitionism, God can do a great, restorative work in your life. Why do you think so much detail about her life is recorded in scripture?; to show forth the vindicating work of the Lord.

Your marriage does not have to be for worse, it can be for better, just like hers turned out to be. Your children don't have to grow up doomed to failure, they can become great and have an awesome impact on the world. So take heart, God will avenge you, by doing right by you. Ask Him to bless your marriage. Trust him to bless your children. Honor Him by living life to the fullest. He will help you. He will restore you. God is faithful. You can choose, for better or…

"Heal me, O Lord, and I shall be healed; save me, and I shall be saved: for thou art my praise." *Jeremiah 17:14*

"For I know the thoughts that I think toward you, saith the Lord, thoughts of peace and not of evil, to give you an expected end. Then shall ye call upon me, and ye shall go and pray unto me, and I will hearken unto you. And ye shall seek me, and find me, when ye shall search for me with all of your heart. And I will be found of you, saith the Lord: and I will turn away your captivity…"

Jeremiah 29:11-14

Journal Entry About the Lord's Favor & Vengeance:

Family & Friends

Bathsheba's story undoubtedly stirred a number of emotions. It should have, hopefully, dispelled the myth that a woman 'has it coming to her', or that 'she asked for it'.

There is no excuse for someone to commit a sexual crime. It doesn't matter if the person flirted, complimented or even admired him to the point of humiliation, no one deserves to be attacked.

As a loved one, you know the victim better than the judge, jury, lawyer or police. They want to know what attire or behavior that may have somehow invited the attacker's advances or if the person was simply in the wrong place at the wrong time. They ask questions that cast doubt because they don't know the victim.

You, however, know who he or she really is. Just because someone is fun loving, friendly and perhaps a social butterfly, does not constitute an invitation for a violent attack. No one deserves that. No one.

Please assure your dear one of this again and again. No matter what they did or failed to do, the only one guilty of a crime is the rapist.

Being friendly or attractive is not bad. As their friend or family member, come to their defense when the enemy tries to discourage and discredit their worth.

Thank you family for:
Helping without humiliating,
Correcting without criticizing,
Teaching without traumatizing,
Mentoring without monopolizing.

Behavior does not equal worth. *"But God commendeth his love toward us, in that, while we were yet sinners, Christ died for us." Romans 5:8.* Don't let your dear loved one feel like she or he deserved that assault. Don't let the authorities try to mortify them into some sort of recant.

God's purpose for them is greater than the enemy's attempt to thwart it. Even if they are married to the very one that defiled them, still, God's hand is on their lives. Let Him work this out. He did it for Bathsheba. "...all things work together for good to them that love God..." Romans 8:28.

section three

"...to set at liberty them that are bruised..."

"And it came to pass in those days, when there was no king in Israel, that there was a certain Levite sojourning on the side of mount Ephraim, who took to him a concubine out of Bethlehem-judah. And his concubine played the whore against him, and went away from him unto her father's house to Bethlehem-judah, and was there four whole months. And her husband arose, and went after her, to speak friendly unto her, and to bring her again, having his servant with him, and a couple of asses: and she brought him into her father's house: and when the father of the damsel saw him, he rejoiced to meet him. And his father in law, the damsel's father, retained him; and he abode with him three days: so they did eat and drink, and lodged there.

And it came to pass on the fourth day, when they arose early in the morning, that he rose up to depart: and the damsel's father said unto his son in law, Comfort thine heart with a morsel of bread, and afterward go your way. And they sat down, and did eat and drink both of them together, for the damsel's father had said unto the man, Be content, I pray thee, and tarry all night, and let thine heart be merry. And when the man rose up to depart, his father in law urged him: therefore he lodged there again. And he arose early in the morning on the fifth day to depart: and the damsel's father said, Comfort thine heart, I pray thee. And they tarried until afternoon, and they did eat both of them. And when the man rose up to depart, he, and his concubine, and his servant, his father in law, the damsel's father, said unto him, Behold, now the day draweth toward evening, I pray you tarry all night: behold, the day groweth to an end, lodge here, that thine heart may be merry; and to morrow get you early on your way, that thou mayest go home.

But the man would not tarry that night, but he rose up and departed, and came over against Jebus, which is Jerusalem; and there were with him two asses saddled, his concubine also was with him. And when they were by Jebus, the day was far spent; and the servant said unto his master, Come, I pray thee, and let us turn in into this city of the Jebusites, and lodge in it. And his master said unto him, We will not turn aside hither into the city of a stranger, that is not of the children of Israel; we will pass over to Gibeah. And he said unto his servant, Come, and let us draw near to one of these places to lodge all night, in Gibeah, or in Ramah. And they passed on and went their way; and the sun went down upon them when they were by Gibeah, which belongeth to Benjamin. And they turned aside thither, to go in and to lodge in Gibeah: and when he went in, he sat him down in a street of the city: for there was no man that took them into his house to lodging.

And, behold, there came an old man from his work out of the field at even, which was also of mount Ephraim; and he sojourned in Gibeah: but the men of the place were Benjamites. And when he had lifted up his eyes, he saw a wayfaring man in the street of the city: and the old man said, Whither goest thou? And whence comest thou? And he said unto him, We are passing from Bethlehem-judah toward the side of mount Ephraim; from thence am I: and I went to Bethlehem-judah, but I am now going to the house of the Lord; and there is no man that receiveth me to house. Yet there is both straw and provender for our asses; and there is bread and wine also for me, and for thy handmaid, and for the young man which is with thy servants; there is no want of anything. And the old man said, Peace be with thee; howsoever let all thy wants lie upon me; only lodge not in the street. So he brought him into his house, and gave provender unto the asses: and they washed their feet, and did eat and drink.

Now as they were making their hearts merry, behold, the men of the city, certain sons of Belial, beset the house round about, and beat at the door, and spake to the master of the house, the old man, saying, Bring forth the man that came into thine house, that we may know him. And the man, the master of the house, went out unto them, and said unto them, Nay, my brethren, nay, I pray you, do not so wickedly; seeing that this man is come into mine house, do not this folly. Behold, here is my daughter a maiden, and his concubine; them I will bring out now, and humble ye them, and do with them what seemeth good unto you: but unto this man do not so a vile thing. But the men would not hearken to him: so the man took his concubine, and brought her forth unto them; and they knew her, and abused her all the night until morning: and when the day began to spring, they let her go. Then came the woman in the dawning of the day, and fell down at the door of the man's house where her lord was, till the light. And her lord rose up in the morning, and opened the doors of the house, and went out to go his way: and, behold, the woman his concubine was fallen down at the door of the house, and her hands were upon the threshold. And he said unto her, Up, and let us be going. But none answered. Then the man took her up upon an ass, and the man rose up, and gat him unto his place."

Judges 19:1-28

Chapter Seven: Twice Shamed

I don't think I need to state that this account in scripture is undisputedly one of the most heart-wrenching stories ever recorded. It encompasses a number of elements that are so relevant today that I am not sure how to approach this. The story starts off showing great aspects of human nature such as; kindness, hospitality, communion and so forth. It ends up showing some of the worst evil expressions that humans are capable of committing.

The text tells us that after the Levite went to this woman, spoke 'friendly' to her to convince her to come back to him, that he then sat down and dined with her father in fellowship again and again—possibly as a show of good faith of his character. One would think that this man really cared for her. Yet, the first chance he gets to protect her, he forgets who he is and uses her to protect himself. He wouldn't lay his life down for her, but he literally forced her to lay hers down for him.

What about the old man? He seemed to be so polite at first, opening up his home in such a warm and hospitable way. It was bad enough that he offered his own daughter to the men, but how is it that the he felt he had the right to offer the concubine of his guest to those beasts (men) as well? Why wasn't the servant sent out to appease them, or was he of more value just because he happened to be a man? Why did the Levite take it upon himself to save himself and his servant and offer his concubine? We may never fully know the value systems of people back then regarding women, although, in some parts of the world today, we see reflections of this kind of disrespect.

Why is a man always called the husband to both wife and concubine? Why doesn't he have a different type of name to denote his status? What is a husband anyway? Isn't he the one who has pleaded with the family of the woman he says he loves to allow her to his care? Isn't he supposed to be the protector, the provider and the priest of the home?

The three men in that house stayed shut up inside while that dear woman was being raped all night long. Where was their anger, their valor—their decency? It's pretty difficult to tell which of the two groups of men were more wicked. It would have been better for the three of them to have all died trying to save the woman, then to let her be attacked so viciously while they did nothing in her defense. This is how shame comes twice, when the one who should be there for you actually contributes to the pain. (And we all know that for some the source of pain is the very one who should be protecting them.)

Where can one turn when the protector is not protecting? It is clear to me that many of the men of that time did not understand the worth of a soul. Women were regarded as property far too often. It seems almost as if they were used as money or a prize; apparently their souls were not considered any more than one may regard livestock. Thank God all men do not think this way.

Perhaps the old man, the Levite and the servant were of the opinion that the men would not hurt the women; this could explain the old man's decision to offer his own child. But, if that is the case, couldn't they hear the woman's cries for help? It's hard to figure out what these men were thinking.

Then again, it's hard to understand many men now. Some still 'push' the women in their lives away when they no longer feel they have use for them. Divorce is so common, it's almost inconceivable to imagine the time when it was rare.

What was it that drove those men to put that dear soul out there that night? I can't stop asking that question. Was it cowardice or just a general disrespect for women? Maybe it was sheer ignorance. For all intents and purposes, they may have reasoned that a woman's body was essentially designed to be for a man. Perhaps by culture they thought that offering females to those sodomites would stay their ferocity by the natural softness and physical attributes of a woman.

What they apparently did not understand, is that a woman is not just a warm body. It is the spirit of the woman residing in that body that is significant. She is a living, thinking, destined person who was created on purpose by the only wise God, who never wastes His creative efforts on the mere satisfaction of man.

Ignorance is nothing new. Many people still predominantly esteem women for their physical attributes before really making an effort to know them. So apparent is this that some cultures force women to cover their entire bodies from head to toe in quite uncomfortable and binding ways. In sharp contrast, other cultures encourage women to "liberate" themselves by exposing themselves in exhibitionistic ways to feel empowered.

Marilyn Monroe was one of the most photographed of women in entertainment history. Her conspicuous persona and beauty were admired and almost worshipped by many, even decades after her death. Yet, the ones who knew her personally spoke of a very different person than the overly sensual person she publicly portrayed herself to be. One of her acquaintances even called her "…one of the saddest women he ever met." Imagine that. There was more to her. There is more to all of us. It is one's entire being that matters-- spirit, soul and body. For a man to regard a woman merely for what she can do to satisfy him is wicked and reprehensible. However, when a man does treat a woman with disrespect, he does so because of the enemy's hateful influence regarding women.

"*…and I will put enmity between thee and the woman, and between thy seed and her seed…*"

Genesis 3:15

As we can see here in this particular part of the curse on all mankind, women suffer grievously from the enemy's distinctive and direct hatred towards women. As a result, women throughout time have suffered grave indignities and cruel disrespect. Thank God that every aspect of the curse was handled on the cross. Jesus made a way for all who believe to be free from the curse. (See Galatians 3:15) No true Christian man will treat a woman with disrespect and contempt.

How God really wants women to be regarded is quite evident by the command that Christ set: "*husbands, love your wives, even as Christ also loved the church, and gave himself for it...*" (Ephesians 5:25). A real man of God would likewise lay his life down for his wife. It is a shame on both sides when this is not the case. Shame on the man for his lack of true Christian masculinity and shame for that woman who has to endure living in subjection to his carnal indifference.

"*God is not willing that anyone should perish...*" (II Peter 3:9). His heart longs for all of us to do well. The Lord would never do to us what that Levite and the old man did to those women. He would never put us is in harm's way to protect Himself. Neither would he ever condone such behavior. I cannot imagine what the old man's daughter must have thought of him after that night. The shame of living in a community where such violence prevailed was not comparable to the dishonor of knowing how her own father considered her. Regardless of the culture of that time, she had to be mortified and greatly hurt, to say the least.

It was probably the same contempt that Lot's daughters must have felt when he offered them in the same manner centuries before. (See Genesis 19:1-8) Undoubtedly those girls no longer had respect for their father. ("*...a brother offended is harder to be won than a strong city...*" Proverbs 18:19) They became so unbalanced that they made him drunk in order to sleep with him, thinking it was the only way to preserve the bloodline. This demonstrates all too clearly that warped behavior in adults often produces warped mentalities in children. (See I Corinthians 15:33 Jeremiah 4:22)

It needs to be stated, in retrospect, that the dear woman that died before the behavior of those Benjamite men was stopped did not die in vain. Before her death we see in previous scriptures, that women were offered in the same way to ravenous men as in Lot's case, as well as when they were possibly treated likewise when they were in Egypt, as was mentioned in this text (Judges 19:30).

No where else in scripture is it recorded that women were ever again voluntarily given over to men in that manner after her death. To this day, though women are still taken as sex slaves, children used for pornography, boys assaulted to avoid pregnancy and countless women are punished with rape by religious zealots in the name of honor, there is still hope. Now there are laws in place around the world to check this perverse and atrocious behavior.

The rest of the story that follows our text tells us how the concubine's husband did something unheard of after her death. He divided her corpse into 12 parts and sent them to the twelve tribes of the family of God's chosen people, the Israelites. Appalled, they demanded to know the meaning. The man told them what happened and needless to say, a war raged to stop this behavior as each tribe sent valiant men to halt the city where such a horrible crime was openly committed.

I don't know what went through the mind of the dying woman as she found her way back to the house where her husband was staying. I doubt she knew that thousands of men would lay their lives down as a result of her death. Did she know that she was not worthless? Was she able to realize that she was precious regardless of how she was treated in her last hours?

Perhaps her husband, from all appearances in the text, didn't expect her to die. In his mind he may have thought to punish her for her whoredom. Maybe he feared his own life; or perhaps he thought that the Lord would miraculously rescue her with an angelic intervention, as was the case with Lot, we can't really be sure. One thing we are certain of, that young woman, who is nameless to us, was precious to God. So much so, He saw to it that her story was recorded for all history.

God allowed not only her death to be avenged, but also her honor. The battle was to requite her death. To vindicate her honor, the remaining tribes vowed not to give their daughters in marriage to the Benjamite tribe because they refused to punish the men who ravaged her so relentlessly.

The sadistic behavior that those men exhibited had to be stayed. Why so much trouble for one woman? For the same reason that I am sitting up right now in the middle of the night penning this text…every soul matters to God. God is *"not willing that any should perish…"* but that all men might be saved.

This story didn't quite end there. After the people's quest for justice and communal purging was satiated, there remained the issue of the vow they all declared concerning giving their daughters in marriage to the Benjamites. Ultimately, this would mean that the Benjamite tribe would slowly come to an end, having no posterity. They got around it in an interesting way. You can read about it in Judges 21. As I have emphasized before, God is not willing that any should perish.

"The Lord is not slack concerning His promise, as some men count slackness; but is longsuffering to us-ward, not willing that any should perish, but that all should come to repentance."

2 Peter 3:9

Many of you may not really relate to such an extreme situation as the woman we are reading about, but no doubt you do understand the shame that seems to cloud over a life after a sexual crime has been committed. The feeling becomes worse when a loved one contributes to it. There can be a strong tendency to take on another personality entirely because of the weight of disgrace. Don't allow what happened to you, or anyone who knows about what happened, make you into someone that God never intended you to be. You are more.

I remember walking from the mall near my home one beautiful day several years ago. I had just enjoyed a great time at the gym and was looking forward to my plans for that night. Right then, I ran into an old High School teacher of mine, one I recalled as being the one who believed in me, applauded my leadership skills and was instrumental in getting me involved in many wonderful opportunities to advance my aspirations. I remember how excited I was to see her again as I approached her.

Like all of my teachers, she looked exactly the same to me even though nearly ten years had passed. She was thrilled to see me as well, for a moment. Then she put the face with the memory of what happened to me and her countenance changed. Her pity did nothing but make me feel shame- - again. "Yes, I am the one...", was all I could answer when she inquired about the incident. I ended the conversation as politely as I could, trying my best to hide my shock and humiliation. I was not at all willing to linger in that pity party, even if it was for me.

I walked on in complete disbelief. After all of the things that God allowed me to accomplish in High School, participating in extracurricular programs, experiencing many academic accomplishments, receiving several awards as well as being involved in the community and graduating with honors. All of this, in spite of all that happened to me, did not seem to register with this teacher at all. She seemed to recall only that I had been victimized.

I felt so astonished when I left her. I was grieved, ashamed and disappointed. Yet, while all these emotions washed over me like a flood, I was able to remember something. I am more than the remains of what happened to me. There was no way I was going to allow anyone to treat me like a pitiful victim. Jesus said that " ...a *man's life consisteth not in the abundance of the things which he possesses." (Luke 12:15)* I may have possessed memories of experiences, some heavy emotion and even dreadful thoughts from time to time, but that wasn't what I was. I am more than the result of my experiences—and so are you. If I saw that same teacher right now today, I would probably respond to her memory of me with a smile. I am not the wounded child she knew so long ago. I am free.

As long as we live on this earth we will endure some hardship, some sorrow and some level of pain. This is the common fact of life shared by all mankind throughout the world and all time. The only difference is the fact that some have been blessed to learn who to go to for help.

This is why it is so important to embrace the truth of God's Word. He has provided it to help us. We need to learn and understand who He is so that we can move forward into the destiny we were born to know. In every aspect of our lives, good or bad, the presence of a loving Father makes a tremendous difference. Though this blessing may not be a reality (in the natural human sense) for everyone, it is still true. God gave us fathers because we need them.

Essentially, God is the Father of life. He is the greatest father of all fathers and His love is matchless. He stands ready to guide, instruct, correct, care for, protect, cherish and yes, comfort us. But He only fathers those who by faith allow Him to. *"Behold, I stand at the door, and knock: if any man hear my voice, and open the door, I will come in to him, and will sup with him and he with me."* (Revelation 3:20) God longs to embrace His children, blanketing them gingerly with His deep affection for us all.

We must learn to accept the love the Lord God extends to us, for He is quite demonstrative with it if we choose to notice. Creation shows off the importance of God's attention to detail. He made the entire universe painstakingly with great thought and high regard. Not one snowflake is like any other and the waves of the sea never touch the shore in the exact same pattern.

The birds sing and the sun sets in a marvelous display of unpredictable beauty day by day. How can we look at such glorious splendor and not marvel about the measureless beauty within the Creator of all that is good?

Think about it. Would this great God of the entire universe, who made everything fit in place so wonderfully really find no interest in you? Of course not. God thinks of you constantly. His thoughts for you are more numerous than every single grain of sand on every shore of every coast- -combined! *"How precious also are thy thoughts unto me, O God! If I should count them, they are more in number than the sand: when I awake, I am still with thee."* (Psalm 139:17-18) With all His thoughtfulness, the Lord would never think of forcing you to follow Him--even though He could.

The Lord wants us to discover Him willingly. Love is most vibrant when it is equally exchanged. God made the first step when He gave us His only begotten Son, Jesus, the Christ. Anyone who believes on Him will not perish, but have everlasting life. Try to see His love for you. Your freedom from all your hurts and fears is found in Him. *"He is love and "perfect love casts out fear..."* (See I John 4:18). This is what you can expect when you come to the Lord with your whole heart. Don't hold anything back from Him.

Search the scriptures for yourself. You will find the answer to all of your questions and find out what the Lord is saying to you. Once you begin to understand God's heart, You will begin to *"...know the truth and the truth will make you free."* (See John 8:31-32) Jesus is the truth. Freedom is what you need. Freedom from the fear, the guilt, the sorrow and the hate. Freedom from every level of pain. Freedom from shame.

It's time for you to be you again. God wants you to experience true liberty in your soul. You don't have to live the rest of your life still grieving for what was stolen from you. The suffering ends the moment you trust God to make you whole again. Let Him do it. Allow the Lord to love you, father you and restore you. Then you will be able to stand erect, with your countenance lifted and your heart strong. No one will be able to make you feel mortified again. They won't think to define you as a victim because you are not one.

Others may go on to say: "Well, you asked for it! ;" "I knew you would get yourself into trouble;" "You're tough, get over it!;" "Don't you dare say such a thing about him!" Or the old: "You probably brought this on yourself." You cannot believe these witnesses. Some people in their ignorance may be insensitive enough to blame you. They are not able to support you because their opinion is one-sided. Forgive them, dismiss their behavior and disregard their false words.

It's hard enough to live with the trauma of what happened to you, but to endure the shame twice, the second blow being from your loved ones and acquaintances, can almost equal in agony. Now, they may not say it to your face or say it at all, but sometimes you know when you look in their eyes, that they can't see you any other way. Other times a hint of disbelief can be caught in their facial expression. Even a sigh can be tough to endure. But you know what? What they believe or find hard to believe is not important. They can't heal you anyway. You must learn to rely on the one who has your heart in His hands.

The fact is, everyone is not going to be able to wisely handle what happened to you. You will have to find freedom in God's truth because you cannot rely on their strength to hold you up. One truth you must focus on continually is that only one individual truly understands you, the Lord. He will pick you up and turn everything around for you. Without relying on Him, you may find yourself set up for a tougher reality if you try to rely on people alone. Why? Because if you are waiting on loved ones to be strong, you will be disappointed. They really aren't able to be strong because they have their own pain to sort out. Only God can restore you and bring total closure to your pain…and theirs.

True enough, family holds a stronger connection to you than the one who assaulted you. He was only able to abuse your body and wound your soul, but your loved ones can break your heart if they are not considerate as you recover. As much as we all would like to believe that our dear ones will be there for us, it isn't realistic or even fair to expect them to. We have to turn to the Lord, the one who knows how deep the hurt cuts, and let Him be strong for us. Family can be supportive, but at times, they need a hug from you as much as you need it from them.

"Draw nigh unto my soul, and redeem it: deliver me because of mine enemies. Thou hast known my reproach, and my shame, and my dishonor: mine adversaries are all before thee. Reproach hath broken my heart; and I am full of heaviness: and I looked for some to take pity, but there was none; and for comforters, but I found none." *Psalm 69:18-20*

How can you deal with the awkward stares and long stretches of silence from your well-meaning loved ones? Prayerfully. Whisper to the Lord a prayer for peace whenever those who 'know' make you feel shame all over again. At times you won't even be thinking about what happened, and someone will walk in and say something insensitive-or worse, say nothing at all. Then you realize that they no longer see you the same. You become uncomfortable and somehow exposed all over again. In all honesty, and it is a flaw in our human nature, family simply cannot always be there. They don't know how to be.

One day, you will be able to share your freedom with them and they will find comfort in just knowing that you are whole. The more you express to them about the goodness of God, the healer of broken hearts, the more they will want to be healed, for their hearts have suffered greatly in all of this as well as you.

The Israelites handled the situation concerning the concubine in the only way they knew how, with violence. But this is a new day and there is a stronger way to handle this kind of 'situation' with a better method of force. That force is fervent prayer. Prayer that is purposely directed and founded in the written Word of God, not in exasperated emotion and complaint. To just pray for comfort and consolation is fine when you are first recovering, but there comes a time when you have to pray for total restoration. (See Luke 18)

"...the kingdom of heaven suffereth violence, and the violent, take it by force." *Matthew 11:12*

You have to make a stand for who you are. Forget what is common, this is about your destiny. The world is willing to accept you as a victim, but that is not who you are. Many won't mind you ending up getting lost in some type of 'lifestyle' as long as you don't make them uncomfortable. But this is not about their will, this is about you taking your true place to affect the world as you were born to do. Of course you may have dark days and mood swings and bouts of sorrow and confusion. Maybe you won't have one single available friend in the world to call on at the lowest times. You may even want to end it all. That's kind of how it is dear heart, when you've been hurt that deeply.

Please believe me when I tell you, you will not always feel that way. When I think of the verse in scripture that reads, "*a wounded spirit, who can bear?...*" (Proverbs 18:14), I relate immensely. We all do. No human can handle or heal a broken heart; thank the Lord that He does. "H*e healeth the broken in heart, and bindeth up their wounds...*" *Psalm 147: 3*

Are you willing, yet, to let Him heal you? You do not have to carry this pain inside of you day after day and year after year. Some of you have never even voiced your experience, but you know, God still hears the cry of the soul, whether you have wept inside for days or for decades. Let your cry become your prayer. (Psalms 42:11 *"Why art thou cast down, O my soul? And why art thou disquieted within me? Hope thou in God...")*

It comes to this, you have to fight for your soul. I know that the last thing you may want to do is have to face off to anyone, but confrontation is not the directive here. The charge is to put an end to all the agony and grief that resulted from the crime. This thing didn't come to you softly, but fiercely, so you cannot be passive now.

By now it should be very clear to you, that you are special to God and that you have a right to be free to walk out your destiny in peace. Take back what was stolen from you with all the power of heaven

behind you, by prayer. Put on the armor of God daily. (See Ephesians 6: 12-18) Make your mark in this world, scars and all. You have purpose. You are destined. You are full of wonder and life, so live.

If you have to have every Christian you know to pray for you, then put in your requests. If you need to put scriptures all over your walls until you know them- -fine, do it. If you have to get in every prayer line and respond to every altar call-then go! Pray until you are restored, and don't you give up until you are walking in that blessed state of fulfillment that God intended for you.

"The Lord shall cause thine enemies that rise up against thee to be smitten before thy face: they shall come out against thee one way, and flee before thee seven ways." *Deuteronomy 28:7*

Sure, the attempted assassination of your soul began as one sharp blow, but as you come out of this, not only will you be whole, you will be rid of weakness, ignorance, isolation, silence, worry, blindness and passivity. How's *that* for seven ways the enemy will flee from you? For weakness you will have valor. For the areas that you were ignorant, you will be wise and instructive. Instead of being isolated, you will reach out to others and strengthen them. No longer silent, but with a clear, resounding voice of victorious testimony. Worry will no longer be a part of your thought processes, only faith and hope for an even greater journey. Blindness will clarify to perceptivity and foresight. Never passive, but selective of which battles are worth fighting. For this you must press forward, for this you must fight.

"...forgetting those things which are behind, and reaching forth unto those things which are before, I press toward the mark for the

prize of the high calling of God in Christ Jesus."

Philippians 3:13-14

You are strong by virtue of Who stands with you. Allow the Holy Scriptures to direct you on just what to do to recover all that you lost, including your true identity. *"Thy word is a lamp unto my feet, and a light unto my path".* (Psalms 119: 105) You are not a victim. You are precious and you belong to God. Enjoy the love God is extending to you. As I learned from hearing a former Korean Comfort Woman's resolve, "You have no reason to be ashamed, you did nothing shameful, that {criminal} did."

"I love the Lord, because He hath heard my voice and my supplications. Because he hath inclined his ear unto me, therefore will I call upon him as long as I live. The sorrows of death compassed me, and the pains of hell gat hold upon me: I found trouble and sorrow. Then called I upon the name of the Lord; O Lord, I beseech thee, deliver my soul. Gracious is the Lord, and righteous; yea, our God is merciful. The Lord preserveth the simple: I was brought low, and he helped me. Return unto thy rest, O my soul; for the Lord hath dealt bountifully with thee. For thou delivered my soul from death, mine eyes from tears, and my feet from falling. I will walk before the Lord in the land of the living. I believed, therefore have I spoken: I was greatly afflicted..."

Psalm 116:1-10

Chapter Eight – Erase the Trace

Looking again into the story of the Levite's concubine, we have to consider the horrible civil war that resulted as a consequence for what the sodomites did to her. Nearly the entire community of God's people was involved in the effort to rid themselves of the evil that caused the woman's death. They knew they had to erase every trace of this type of mental plague. To allow people to behave as those men did threatened the health and well being of the entire nation. They had to check it, before it reached an epidemic level, and they did so with God's authorization.

Every other occurrence in the Old Testament of lewdness was met with harsh penalty of death, usually by God's own hand or with His insistence and/or assistance. The cities of Sodom and Gomorrah as well as the nation of Israelites that engaged in sexual perversion as they wandered in the wilderness are clear examples. (See I Corinthians 10:7-8) What is it about sexual sin that is so much more vile? In this passage we can see why.

"Flee fornication. Every sin that a man doeth is without the body; but he that committeth fornication sinneth against his own body. What? know ye not that your body is the temple of the Holy Ghost which is in you, which ye have of God, and ye are not your

own? For ye are bought with a price: therefore glorify God in your body, and in your spirit, which are God's." *I Corinthians 6:18-20*

Sexual sin destroys you. You are no longer your true self—your best self. Your conscience is changed whenever you engage in sin, but more so with sexual sins because you aren't just hurting someone else as is the case when you commit any other sin. You are hurting you own body.

Your physical and mental health is affected when you engage in sexual sin. You develop disrespect for your body and carry yourself differently when you are practicing sexual sin. You open yourself up to strange diseases, cultures and disturbing behavior when you connect yourself to someone sexually. You can no more separate yourself from someone you have had sex with than you can separate the river from the ocean it flowed into. Once they are joined, they are one.

The same goes for every, stream, brook, creek and any other body of water. When they mix into the ocean, they can never again be separated. Likewise, when you join yourself to someone sexually, you take inside of you attributes of every other person they were previously joined to. This very truth is why the Israelites had to stop the men from spreading their wantonness. No man, woman or child would have been spared from the threat of the same disgraceful peril as the concubine's, if that mentality was allowed to grow.

I believe with all my heart that the Lord allowed every horrid line of her story to be included in the bible to restrain us for all time, never allowing such a thing to happen again. It may be hard to read, but we need to be careful that history is not repeated by allowing a barbarous wolf to hide in our communities under the sheepskin of liberty or human rights, and consequently open others to harm.

My grandfather, Thomas Murray was a very wise man. He told me many things as I grew up. I remember him telling me that "when a person does something wrong, at that moment he {most likely} believes that he is right;" This explains why so many can commit such heinous crimes and feel absolutely no remorse. Or even why some fight over their choice to pursue unnatural affections like homosexuality as a right, instead of an abominable sin against God.

For all the arguments that are out there, I must say, God alone is always right. We need to follow His certainty and not the feelings of others. Otherwise, we will only sit by and watch our communities become inhumane and maddening. Are we really that far off from the mentality of the Benjamites? Weren't they exercising their 'rights??!' This is about as insane as it gets.

At any rate, my purpose for writing is not to challenge the determination of those engaged in sexual sin, my heart is to help those who have been hurt because of sexual crime; to help erase the trace that the offense left behind. The trace of hatred, bitterness, confusion or worse, transfusion. That is, to become an entirely different person, taking on a strange identity, embracing an unnatural affection or following a darkened path of imitated behavior. This is the trace that needs to be removed.

The Israelites knew the importance of this and waged war against the Benjamites after they refused to turn the guilty men over for punishment. Thousands died on both sides of the battle. In the end, the remaining few Benjamite men weren't going after other men, they married women to keep the family line from annihilation.

This is vital to note, annihilation. Remember in the first chapter when we looked at Jesus description of the enemy's objective- -to kill, steal and destroy? That is what annihilation is. To cause to cease to exist. When men turn to men in unnatural affection, the bloodline is diminished. When women turn to women, the bloodline is discontinued, for only by a man and a women coming together can pro-creation of human life take place. One way or another, a mental change takes place after a sexual sin has occurred that causes aberration. There are even terms that better identify this state:

Emasculate: To deprive of virile or procreative power:

Castrate; to deprive of masculine vigor or spirit: weaken

Effeminate: having feminine qualities (as weakness or softness) inappropriate to a man: not manly in appearance or manner: marked by an unbecoming delicacy or over-refinement.

Masculinize: to give a preponderantly masculine character to esp: to cause a female to take on male characteristics.

Defeminize: To divest of feminine qualities or characteristics.

Why is the enemy of our souls working so deliberately to keep children from being born? To ruin a man or woman by sexual crime to the point where they become homosexual or even anti-sexual, means they may never have children. This is no more than castration of the soul. I cannot say it enough, sexual crime is never just about sex.

Children are precious and every life has purpose. Every new soul that is born into this world and draws in the breath of life is a front to the kingdom of darkness and a threat to the plan of the enemy of God. One method of this enemy is to contradict the basic natural instinct of mankind and make him believe a lie about himself by using unclear and presumptuous arguments to bring about a conclusion that is dense with confusion. Examine the chart on the next page carefully:

The difference between what God says and what man (deceived by Satan) says:

The Truth: God's Word	The Lie: Satan's deception
It is written… (see Luke 4:4)	Hath God Said?
God knew you before you were formed in the belly… (see Jeremiah 1:5)	Are you really the gender you are supposed to be? There must be a mistake… You have the wrong body… it's biologically proven to be genetically and pre-natally determined…
Come out from among them and be separate… (see 2 Corinthians 6:17)	You need to be around people who are just like you…they are the only ones who understand…It isn't sin, it's human biology
Male & Female created He them… (Genesis 1: 27)	Homosexuality is an individual right of choice…Be what you want to be…
The just shall live by faith and not by sight…(Romans 1:17)	Do what you feel inside…it's your body…

The deception can seem pretty convincing because it lends the sensation that a need is actually being met. The fact is, the need is not met nor can it be by the enemy's lie. For example, the hold that effeminism and homosexuality has is very often comfort, false security and the feeling of belonging. To fit in is a strong desire for anyone. To be rejected is one of the greatest fears, at least... for many. Everyone wants to be wanted and accepted, because that is the nature of humans. It appears easier, safer and more or less challenging to either mimic what intimidates us or join ourselves to those who share the same (fearful) challenges.

A man afraid of being rejected as the man he is, can gravitate to behaving like the women that he is patronized by. Observing feminine ways, lingo, style and distinction. Studying them because it is safer emotionally than to make an attempt at masculinity.

A woman who feels overlooked by males can become highly interested in predominantly male interests. Learning the mannerisms, staying in male-dominated hangouts, donning men's attire and speaking in the same manner as men commonly do. This is true in many instances, especially in the young and insecure. The psychosexual seed planted in either male or female produces both devastation and dis-ease. But in a male, it can mushroom. A male has the power to do the same to others, if not by force, by initiating perverted activity with willing parties in a distorted and degrading way. How can a man define himself as a man if the very thing he was always taught that made him a man was defiled, confused and greatly disturbed by rape?

Gender confusion is nothing but an old lie from that same serpent of *"Hath God said…?"* Still he preys on the ignorant and coaxes the arrogant. Eve was deceived by the devil, Adam was not. (See I Timothy 2:14). To this day, the two types of people still exist, those who know better and choose to do wrong, and those who honestly think the lie of homosexuality is a *good* idea. In either case, the wrong is still an abomination to God.

If you read the account of the cities of Sodom & Gomorrah, you will notice that all of the people from old even to the youths came out from every region to participate in the plan to gang rape Lot's visitors. All of them. (Genesis 19:4) That spiritual atmosphere had infected particularly every adult male citizen in those cities. Lot was there temporarily and because of promise, was under God's merciful protection. Although he didn't participate in such behavior, he was still affected morally to the point where he held a disregard and disrespect for women.

Those citizens were blind spiritually and mentally long before the angels struck them with physical blindness. Just as the angels caused them to be unable to see or find the door to where they wanted to go (Lot's house), the enemy had beforehand blinded them to finding the door to where God wanted them to go to become (true men). These men were offered the virgin daughters of Lot by Lot himself.

The men wanted the visiting males (which, unbeknownst to them were angels) to feel what they felt and experience the same enslaving horror. To be emotionally castrated, so to speak, as they had been. The goal was to consummately weaken them for an indefinite time so that they might be taken advantage of further. Who knows? This practice may even have been how they worshipped idols. They may have subjected Lot to the very same ritual, only God knows.

The real hidden goal in all of that was again the work of Satan. He went as unnoticed as a leech, working towards the same type of result. A leech bites its victims in such a stealthy way that it is undetected. It releases into the victim not only a pain-killer but a blood-thinner to make the blood run quickly by stopping it from clotting. Once the leech is full, it releases its grip and leaves the victim, wounded and bleeding. By then the leech is up to ten times its original size.

That is exactly how the enemy wears down thousands into homosexuality. He uses deception to ease the pain they have from feeling different. Then he surrounds them with others with the same type of issues to get them into the 'lifestyle' quickly. Afterwards, he moves on, leaving the person wounded inside with a sorrow that cannot be described and a wound that will not heal—more different and in more pain than they were before. We cannot allow ourselves as vessels of the living God to not at the very least utter a word of prayer for those we see or suspect are in bondage to homosexuality, effeminism, defeminization or emasculation.

By emasculation I mean a male who isn't effeminate, but is not secure in his manhood. He doesn't exert himself as a man. He doesn't walk in the leadership anointing he was born with and is overcome with either fear, an inferiority complex, low-self esteem or even introversion. This is a male who shows little or no signs of assertiveness, ambition, will power, determination or qualities of being self-defending. He exists in a merely surviving, reclusive demeanor, never really striving or reaching very far for anything. He shies away from challenges, retreats from battles he could have won and sets only goals that are miniscule or easily attained. A man who acts powerless.

Virile: Having the nature or qualities of a man. 2. Masterful.

An overbearing parent may emasculate a boy. It is no surprise that many men and youths that behave effeminately were raised by single and sometimes bitter mothers. A man can become emasculated by a father that pushes so hard into macho-ism that he weakens him. He falls into a state where he doesn't even care to try to be a man due to fear of failure. The joy of the 'rite of passage' into adulthood is lost due to the fathers constant provoking. (See Col. 3:21)

On the other side, some mothers can be overprotective, soft, overly sensual or feminine and unreasonably pushy in addressing their sons' individualism. This can interfere in the intricate molding of a young man's mind because the balance of masculinity is not displayed to him. (Proverbs 1:8 speak of the *"...instruction of thy father, and forsake not the law of thy mother."*) A single mother may think she is both mother and father, but she is not. The same is true of a single father, he can never be a mother. This is impossible, only God can be both.

God desires his young men to grow into capable, strong and caring individuals, who are at peace with themselves and possess their own souls. It was never God's intention that a man behave like something he sees just because it seems easier, comfortable or more interesting than what he is. Behaving like a woman may seem more interesting to some boys than behaving like a man. But the will of the Lord is the only way that leads to his destiny. God's wants His men to be man enough to watch over the women and children He gives into their lives, caring for them in every way. Instinctively strong and protective. This cannot be sufficiently developed by the influence of the mother alone, there ought to be a male involved in the upbringing of the child.

"My son, forget not my law; but let thine heart keep my commandments: For length of days, and long life, and peace, shall they add to thee. Let not mercy and truth forsake thee: bind them about thy neck; write them upon the table of thine heart: So shalt thou find favour and good understanding in the sight of God and man." *Proverbs 3:1-4*

God's heart is the same towards girls who strive to arm themselves against (unruly) males by becoming tough and masculine themselves. The Lord did not create the female to be this way and regardless of what has occurred in her life, she does not have to react like that.

Because the devil hates God, he deceives many into thinking they are really not what God intended at all, in order to mock Him. It is indeed that, mockery. When a man turns to another man as he should to a woman, it is mockery, it is disgraceful and it is sin. Among other things, it denies the man one of the most sacred of God's blessings, to become a father. Even deeper still, the deception of the enemy has become so crafty that we could almost call it sophisticated. Specifically, using the results of scientific research to prove that homosexuality is not by choice at all, but a physical or physiological error, thus, indirectly charging that God made a mistake and therefore, cannot possibly be God at all.

This ties right back into why the enemy bothers to disturb people in the first place: to keep them away from the true and living, only wise, Lord God. After all, who would honestly want to serve a God that gave them the wrong body, for no real god would mess up so miserably? Do you see the full circle of the deception? The enemy blinds and confuses people in many ways. Sexuality is only one way. He confuses them right out of believing the word of God that can save them from him.

"But if our gospel be hid, it is hid to them that are lost: In whom the god of this world hath blinded the minds of them which believe not, lest the light of the glorious gospel of Christ, who is the image of God, should shine unto them." *2 Corinthians 4:3-4*

You must come to appreciate the depth of all of this. The point of the enemy confusing people into questioning their sexual 'orientation' (as it is commonly called now), is to have them focus more on themselves than on the plan of God for their lives. As long as we are consuming our energies on ourselves, we end up falling away from what the Lord created us to do. Our concentration becomes wholly invested in what we think we want, not what God wants for us. This pursuit, unfortunately, only leads to the inevitable demise of our very souls. *"For what is a man profited, if he shall gain the whole world, and lose his own soul? Or what shall a man give in exchange for his soul?"* *Matthew 16:26*

We can argue our case and demand our rights, even gathering a following of those who tolerate our belief systems for the cause of liberty. The whole world, as it appears, may even side with our ideals, but at what cost? What is being gained really, if we are not in the defined will of God?

The destroyer's advance has been made. But the reality is this, the devil cannot stop the plan of God. When God wants to bring a life into being, He does so in spite of the attacks on unborn children, abductions of toddlers, pornography, molestation...the list can just go on and on. Thank God the power of darkness is no match for the all-powerful God. *"...but where sin abounded, grace did much more abound."* *Romans 5:20*

This is how this book has come to be, by the grace and wisdom of God, not my brilliance. Even if I had a Doctorate degree in counseling or psychology, I could never come up with the truths that God's word so completely proclaims. I want to make it as clear as possible to anyone willing to consider it, that God has a plan for you that existed before you were a thought in your mother's mind.

If you stop to think about it, why has this horrible occurrence of sexual crime happened to you? Is it that God's plan for your life is so astounding and power-packed that the enemy felt the need to attempt to take you out?; To assassinate your soul?

Maybe, just maybe, there is a special child in your posterity. Could be your great-great grandchild will be graced with the wisdom to break the mystery of mental illness or engineer a mechanism that makes driving safer. This is wonderful to imagine, but if you don't play your part in the quest of life, the whole world could suffer. Everyone matters.

That is why the trace of darkness left behind by the one that harmed you has to be destroyed. If you were born a man, then you are a man. If you were born a woman than that is who you are. Were you outgoing? Then don't become an introvert. Were you the quiet strong type? Then don't be obnoxious and loud to try to demonstrate control. If you were a perfect gentleman before, don't become brutish now. If you were a sophisticated lady, don't turn into anything else.

Whatever you were before you were relentlessly ravaged, you still are. <u>You still are.</u> Who were you? Who are you? Apparently, someone with an awesome, world-changing destiny. Don't let anything or anyone change that. You don't have to. The Lord, He is God. He is willing to save. Let the Word of the only wise and living God completely erase the trace of darkness the enemy used to infect your soul.

"...He maketh me to lie down in green pastures: He leadeth me beside still waters; He restoreth my soul..." *Psalm 23:2-3a*

"...if any man be in Christ, he is a new creature; old things are passed away, behold, all things have become new."

2 Corinthians 5:17

The Lord, He is God. He really is and He is all you need. He made you, and he can make you over again from the inside out. Just ask Him. You don't have to become what you feel you are doomed to be. Just identify the greatness that already resides in who God made you to be. Embrace it. Walk in it. Dance your dance. Sing your song. Breathe.

The Lord is too good to you for you to settle for mere existence in this world. It's time for a change.

"Taletha cumi", which being interpreted means, Damsel, I say unto thee, arise. *Mark 5:41*

This statement from the very lips of Jesus Christ Himself was spoken to a child who was physically dead. Her body stopped working because of illness, but her spirit still lived on. God spoke to her spirit in the same powerful way that He spoke the worlds into existence. His words alone made her body, soul, & spirit come together again, reunite - - resurrect. You have to come alive in the same manner and by the same power. Jesus is still and will forever be... the Christ. He is the resurrection and the life. All who believe in Him, though they may be dead inwardly or outwardly, shall live.

Sexual crime causes a certain kind of death. Death to our peace and security, personality, sense of value and worth. The Lord wants the part of you that came to an end- to begin again. Rape may have killed something deep inside of you, but Jesus is the "...Prince of Life..." (Acts 3:15) You can live again. All you need to do is come to God, and he will transform you. "He that hath the Son hath life; and he that hath not the Son of God hath not life." (I John 5:12) He won't merely change you, because change is revertible. He will make you brand new. Even if you fell into a personality change resulting from the assault, the Lord can make you over again.

"...there is joy among and in the presence of the angels of God over one (especially wicked) person who repents {changes his mind

for the better, heartily amending his ways} with abhorrence of his past sins." *Luke 15:10 (Amplified version)*

The rest of your world may never forgive or accept you once you start again- -and that's not your problem. You have been through enough already, please don't waste your time trying to please people who cannot make you whole. You may have been emasculated as a child and took on the tendencies of a female. Or maybe you were a girl who was hurt so badly by a male or even pressured so much by a lesbian that you took on that identity. You do not have to stay that way, it isn't who you are. *"...let God be true, but every man be a liar..."* *Romans 3:4*

When it comes down to it, it really is just a matter of who you choose to believe. *"...forgetting those things which are behind, and reaching forth unto those things which are before, I press toward the mark for the prize of the high calling of God in Christ Jesus."*
Phil. 3:13-14

All who knew you before only knew you in that darkened state you used to be in. That was the old you. Now God has, through the work of His Son, called you out of darkness and into His marvelous light. You don't have to be effeminate or deffeminate anymore. Through Christ you can do all things. But you must abhor the sin enough to turn from it completely. With the power of the risen Christ inside of you, you will succeed.

It is an amazing thing- - the mind games the enemy puts on people of purpose. When it comes to the issue of sexual misconduct, the plan of the devil is very strategic. Sexual intercourse is meant to be a holy, precious and private occurrence. It is not supposed to be dirty, cheap and on display. God intended it to be an experience of wonder and tenderness between a husband and wife, a chance for one to demonstrate the deepest care and intimacy to another. The connective function of sex is designed to join a man and a woman in an inseparable and permanently spiritual way. It is a bond stronger than that of any other love or natural relationship on earth.

"Drink waters out of thine own cistern, and running waters out of thine own well. Let thy fountains be dispersed abroad, and rivers of waters in the streets. Let them be only thine own, and not strangers' with thee. Let thy fountain be blessed: and rejoice with the wife of thy youth. Let her be as the loving hind and pleasant roe; let her breasts satisfy thee at all times; and be thou ravished always with her love. And why wilt thou, my son, be ravished with a strange woman, and embrace the bosom of a stranger? For the ways of man are before the eyes of the Lord, and he pondereth all his goings." *Proverbs 5: 15-21*

This very real and defining experience cannot be equaled to any other human experience and is sacred in that it was instituted and sealed by the Lord. Therefore, we must conclude that this very awesome and powerful part of life is of great significance. We cannot disrespect its worth by being ignorant of its purpose. This disregard will only result in a chaotic society that hurts itself.

With all that being stated, imagine the impact such a powerful experience could have if it was used against someone. I suppose we could liken it to fire; it is wonderful in that it warms us, helps us to prepare our food and even can be used to protect us and provide light. But with all of its good uses, fire, out of control, is extremely dangerous. Likewise, sex, out of place is just as damaging.

"Can a man take fire in his bosom, and his clothes not be burned?" *Proverbs 6:27*

The plan of the enemy is to persuade us to lose control when it comes to sex. I don't even need to tell you of the perversion that commonly exists today. Understand, there is a plan of action behind the perversion. Consider the following chart:

God's Plan For Sex	The Evil Of Perverted Sex
The sealing of a holy covenant	Without a covenant in place, defilement occurs.
The invisible bond that joins 2 into 1	Multiple partners create multiple levels of confusion, insecurity and disrespectful behavior.
An opportunity to personally minister the fruit of God's spirit (love, kindness, gentleness patience, etc.,) to another soul by covenant.	Lack of covenant causes anger, frustration, distrust and fear.
A tangible metaphor of God's love to us.	Sinful sex has nothing to do with true love.
A gift from God to bless us as individuals.	A trick that only pleases temporarily, but strains indefinitely with guilt, shame, uncertainty and irreverence.
A means to demonstrate a warm, emotional connection to another soul in a cold world.	A crafty tool used to bring about the cold manipulation of another.

One of the most unselfish acts one can do for another, short of giving the gift of life.	One of the most selfish acts that one can do, short of robbing another of everything.
Teaches what it means to be "faithful" as God is by requiring that all others be forsaken.	Causes fear of infidelity and creates hidden suspicion and mistrust via various partners.
A means to strengthen, encourage and affirm another's spirit through comfort, respect and affection.	A means to weaken, discourage, reject and deject another's spirit through manipulation, deception and lust.
The celebration of the possibility of new life and godly posterity.	A subtle means to learn to disregard life and become irresponsible.

When you examine how powerful sex can be, you will come to understand why the enemy perverts it so much. To have such a perverted stain to remain on your conscience can bring stress for years to come. At times, decisions are made entirely because of sexual misconduct from the past.

Get this, if the devil cannot get you to commit sexual sin, he will attempt to afflict you by using someone to commit sexual sin against you, either way, you come out affected. Furthermore, if he cannot get you to commit sexual sin or he finds that he cannot trap you into the net of a rapist, then there is another plan he tries to impose, I call it the "Accusatory Mandate". In this plan, Satan will either prompt someone to accuse you of committing a sexual crime of some degree or he will highlight certain aspects of your personality to have you labeled as sexually damaged. For example, I have observed young boys, (that were gentle, non-athletic or quiet in nature), become labeled as "sissy", "faggot", "soft" and so on. This branding was damaging and defaming. Other males instinctually tended to avoid being identified with them as a self-preservation tactic. In their ignorance, they ostracized the slandered boys whether they believed the lie or not.

In the quest for companionship that all humans have, some of those impressionable boys eventually ventured toward the only ones who befriended them...immature girls. Then by communication and of course by reason of association, they became effeminate, comfortable intimidators and ultimately, driven by the confusion of the devil, homosexual.

Many times males like these may have never been physically molested in any way, but the mental torment was just as disintegrating. They grow up in this life completely isolated by regular men, but embraced by women, who are naturally inclined to draw near to a man who is open and safe.

But I must tell you, this trace of sin, which is part of the plan the enemy has to lock you into a trap God never intended, can be washed away from you completely. You don't have to live up to the expectation of those who merely pity you. You were created by God to be a man, so be a man, even if you were labeled otherwise most of your natural life. Stop gravitating to the women and girls, even if they make you feel 'good'.

Pray for God to deliver you and make your life over again. He can erase the trace. He can remove every bit of effeminism. From the quick-paced, practiced tone of your voice, to the sway in your walk, to the unisex flair of your clothes. God can change you into what you ought to be so that you can accomplish what you were born here to do. God knows you were not put on this earth to be an amateur actor performing day by day, costumed as a woman.

Forget the terminology and arguments of today. This is not about civil rights. This is about peace in your heart. Real peace, not concession. This is about who you are, not what you feel like you are. A man will never know what it feels like to be a woman no matter what he does to himself. Neither will a woman ever feel like a man. Or a bird or a plant or anything else. How can you feel like what you are not? It's all a lie that wastes your time and your energy.

If you were to write down everything you ever wanted to be and everything you ever wanted to do on a sheet of paper, you would have quite a list. But if you were to erase all of the aspirations that were linked to your sexual identity struggles, that list would shorten significantly. Why? Because they are the very lies the enemy has been feeding you for years. Did you really want to be the class clown, making everyone laugh or even stare at you because of the outfit/costume you displayed? Was that your real desire or did you feel you had to make a statement? Who were you living for? What point were you attempting to make? Who were you trying to please?

If you were trying to please God, you missed it. If you were trying to please yourself or anyone else, well, you know the answer to that: wrong again. We were created for God's pleasure and His alone. Would it please God for a man to tell Him he will not be what God wants, but what makes him "feel" good? Of course not. No amount of human reason, intellect or explanation can ever replace God's irrefutable wisdom. It's time to be all that God intended you to be. Erase the trace.

*In churches the youth leaders, ministers, deacons and all those of influence, should pray to receive kind boldness to minister to the young people who are slipping into gender confusion. The sooner it is attacked, the easier the transformation and the less damage the youth experiences. This is my prayer for you:

Lord God, you are the Keeper of hearts and I worship you. You are the Alpha and Omega, the beginning and the end. Open the hearts of the strong men and women of God in the body of Christ. Teach them to be compassionate and willing to receive and disciple people who have been caught in the web of homosexuality but are now free through Christ Jesus. Lord make them sensitive, thorough, diligent, thoughtful, quick to hear, slow to speak and ready to follow the prompting and leading of your Holy Spirit. Work a miraculous wonder in these souls so that even the young boys who were effeminate and the young girls who were masculinized will be embraced, mentored, discipled and taught.

I pray for the binding of the enticing spirits of perversion and lust. Let there be no interruption or opposition from dark enemy forces. Let your work be done in the beauty of your holiness, that these souls will be free from the enemy's snare and the trace of their past ways be erased. In the name of Jesus, the Christ, amen.

Chapter Nine: "You Have No Power Over Me"

The Levite's Concubine was a person that he did not value. Whether it was because of her past indiscretions or just because of the frame of mind that men held in that day and time, for whatever reason, she was just not important to him. In contrast, her father seemed to adore her and want the best for her. He apparently wanted her to stay with him, regardless of what she had done. He was willing not only to look after and protect her, but he also welcomed her husband into his home.

If her husband's treatment of her on her last day was any indication of how he treated her before she left him, it is quite understandable that she ended up with her father again. Not making excuse for her wrong, but we can't help but wonder what her life was like with her husband, as no amount of wrong can justify his disregard for her safety. To give her away to strangers who were dangerous and sick-minded was despicable to say the least.

So, just by observance of his behavior, let's assume there wasn't much love in that marriage. Turning to infidelity and no doubt finding even more disgrace, the concubine went back to where she knew she was wanted, cherished and adored. She went home to her father. No amount of disappointment could take her dad's love away.

She knew who really loved her. In her father's house there was no indifference or disrespect, she was a person, not a possession. It is not inconceivable to speculate that she deliberately committed sexual sin to get out of her marriage and be able to go back to her father. Her father was a kind man.

When her husband gave her up to sodomites soon after his romantic reconciliation, the terror she endured throughout that night cannot be imagined. To be gang-raped and abused for hours at the bidding of one's own spouse is unthinkable. Being at the mercy of sodomites, who showed no mercy whatsoever, was no doubt the worse moment in the Concubine's life.

However, the love she knew from her father was a strong, accepting and gracious love. The kind of love that is powerful enough to make you forget your trouble (*"...love is as strong as death..."* Song of Solomon 8:6). I have learned, by personal experience and that of others, that there is a level of numbing that comes over a person when they are in a continual, binding situation that they cannot get out of. I believe it is a survival instinct that causes us to turn our focus inward when we cannot change what is happening outwardly. Like the singing slave who sings because he knows that his master can never imprison his spirit. (*"The spirit of man is the candle of the Lord, searching all the inward parts of the belly..."* Proverbs 20:27)

If you have ever been in a situation where you are being assaulted continually and painfully, then you know that after a while you find ways to get your mind off of it just to keep from going mad. Severe illness can push a person this way. I would like to think that this woman might have been able to numb herself, to a certain extent, to what was happening to her. Perhaps she thought of surviving and being able to see her father again.

Why elaborate on her dad so much? My hope is that it will become clear that the rape did not erase all the goodness she knew, nor could it. As horrible as it was, and it was just awful by every imaginable description, it did not define her. If we focus on the entire story, we can appreciate the fact that her story is one that demonstrates the significance of a father's love. He did not disown her when she sinned. He did not berate her husband when he came pleading for her to return to him. His kindness speaks of a greater love that extends from the Father of us all. This story is one of preciousness and value. It shows the human spirit in one of its greatest moments: forgiveness; even though it later also shows the human soul at its worse.

Most likely, her father knew the kind of man that his daughter married. Yet he spent time with him anyway, perhaps attempting to impart some level of decency into him or even teaching him a more excellent way to be a man. We can even safely assume that the father detained the man for no other reason than to keep his precious daughter with him even if it meant having her husband living there as well. Truly her father cherished her.

We can't read a story like that and come away with just the horror of how the woman died. We must take into sincere consideration the wonderful joy she knew while she lived. This account is exclaiming loud and clear that sexual crime is not all that there is to recall about someone. Here was a woman, so intriguing that even when she sinned against her husband, he still wanted her. Other men would have divorced her or worse. She was so precious to her father that he let her stay with him regardless of any disgrace he may have encountered because of her sin. Other fathers may have disowned her after the behavior she demonstrated. Her father even respected her enough to allow her to honorably rejoin herself to her husband.

When we first read about her, we may think that she was a victim, but she was so much more than that. Actually, what she really was more importantly, was a daughter. No one can ever take that away from her. Her father loved her perfectly. His care for her was extraordinary. She could always come to him, knowing he would never turn her away. Because of his love for her, no rapist could have any real power over her. All she had to do is hide herself in her father's love, even just to think of it, and she was emotionally and mentally miles away from where she was physically. *"{Love} Charity beareth all things, believeth all things, hopeth all things, endureth all things. Charity* {Love} *never faileth."*

I Corinthians 13: 7-8

This is the kind of love that our heavenly Father gives to us. He takes us in and loves us, cares for us and shows us respect. Once we embrace his heart towards us, I don't think anyone can ever do anything to us that can keep us down. When one is ravaged relentlessly and left to die is such a terrifying occurrence, it is only a moment. It is the whole life, all of it with its' goodness and hardness that make the defining of a life remarkable.

Who were you before the assault? Who are you now? Who were you planning to be before that 'moment' took place? Be clear on this much, for this is what is essential to you moving past the moment and ultimately, being free of it. God's love for us is not unlike that of the woman's father, only, it is stronger and it is greater.

The concubine woman was left for dead. Before the sun rose, she was already gone. Sadly this is not the only occurrence of such a heinous type of crime. The only consolation is that perhaps, just perhaps, the person who was attacked experienced a grace to be able to disconnect mentally from what was physically happening, and find an emotional place of peace. Fond memories, the face of a loved one, the thought of escape, even pity for the attacker himself. Who can know the frame of mind of each individual person? The soul has a magnificent force in willpower. *("He that is slow to anger is better than the mighty; and <u>he that ruleth his spirit</u> than he that taketh a city..." Proverbs 16:32)*

Yet, not everyone is able to disassociate and mentally anesthetize themselves from the madness at hand. For them, death is very different than what the concubine may have known. It becomes an ongoing death of ideals. Dreams, hopes, optimism --all are slaughtered in the massacre of the victim's ease. But death, even emotional death, is no match for the all-powerful God.

Jesus is the resurrection and the life. It doesn't matter if everything you once knew seems to be strange and tainted. Resurrect means to bring back to life. No situation, no pain, no crisis is strong enough to leave permanent devastation because Jesus is our help.

Much of what happens after the assault is still very mental. What were once natural goings and comings of life are now major decisions. Your own home can become a prison because of the very different way you now think. Jesus can make you free from this confining blindness. He alone can make you able to see what you have lost sight of. Free you to be you and let you live again.

It starts with getting to know His words. What He wrote to us is exactly what we need. You will find all the answers to your greatest concerns. Your hopes will be restored and you will be amazed at how common your situation is. There is hope and peace and love and joy for you. God really cares for you. He wants you to become everything He ever intended for you to be. You must take his hand, you see, just choose to believe that He is all that He says He is. Find out for yourself that He is the way. He is the Life you seek. He is all you need. *"...Jesus saith unto him, I am the way, the truth and the life..."* *John 14:6*

No situation has enough power to completely destroy you. It may injure you, but it cannot destroy the real you. Only the Lord has the power to give your life back to you. The greatest alterations take place at His altar. Don't let your life be wasted away by the evils of this world. As long as you are breathing, you are needed in this world. You cannot let everything you once were become a distant memory, felled by a predator's blow. You have access to the greatest source of power that exists. *"...ask, and it shall be given you; seek, and ye shall find, knock, and it shall be opened to you."* (Luke 10:10) What you are really after is not the power, but the one who is all-powerful, that is, the one true and living God, the Lord. As long as you are trusting in Him in faith, nothing will defeat you. No one can possess you. You will not be overcome. No man will ever have power over you- - not the real you. So live your life to the fullest- -from the inside out.

"And fear not them which kill the body, but are not able to kill the soul..." *Matthew 10:28*

I suppose you may wonder why I write so strongly about this when it seemed that the woman in this account suffered unimaginably and then died alone. It seems as if she was conquered- - but was she? Because God is the loving and kind God that He is, I am certain that she was not defeated. In her case, I believe that the Lord took her into eternal rest. Only God knew what she was capable of handling and the events that caused her death were such a burden that she probably would not have been able to live in peace. The Savior of the world, Jesus, had not yet come and hope was hard to grasp. The world was still waiting for the wonderful, counselor called the Prince of Peace. It is possible that God blessed her with an early rest. Who can know?

Today, now that the Christ has come, we have power to overcome any situation because of Him. Jesus came that we might have life and have it more abundantly, nothing can defeat us if we are in Him. We are not hopeless anymore. When we are overwhelmed, we can go to the Savior of the world. If we are hurt inside in a way so deep that only God can heal us, then because Jesus made a way for us, we can *"...come boldly unto the throne of grace, that we may obtain mercy, and find grace to help in time of need."* *Hebrews 4:16*

Whatever the answer may be as to why that dear woman died lies with the Lord. What we do know is, He didn't want her to be forgotten. If He cared so much for her, even to keeping her identity private, I know He is showing care for you right now. No one even has to know that you are reading this book. You can receive counsel, encouragement and true, healing closure without ever having to sit in a group session or call a hotline. Those venues are great, but many will never use them. I believe the Lord provided this resource for all that silently suffer from the pain of sexual sin. Take heart, God knows just how to get to us, doesn't He?

"For God hath not given us the spirit of fear; but of power, and of love, and of a sound mind"... 2 Timothy 1:7

How many people who have lived a life that is enclosed by fear are out there? God only knows. Fear is a dungeon that holds its prisoners captive without mercy or any hope of release. The concubine we have been learning about died after that one horrific night. Many live with the reality of sexual abuse daily.

There are little children that live in homes where their own fathers, guardians or relatives violate them regularly. Some of these men use religion to dominate their women, forcing them to marry as children and even use their position as fathers to 'teach' the girls how to please their future husbands.

Hundreds go through nights of muffled cries and burning tears-children who lie suffocating under the crushing weight of full-grown men who are supposed to be protecting them. They long for the day when they can grow up and be free.

Then there are those who are incarcerated with no hope of escape. Some know they will not be free. Rape is so common that the very thought of spending time in jail brings the subject to mind. It is well known that rape by fellow inmates or even prison guards is a constant evil in male and female facilities, both adult and juvenile.

How horrendous to live with this pressure day after day. How utterly mind- numbing. I could not begin to imagine the intensity of such woe; my mind cannot fathom it. There are many more souls that find themselves caught up in strange rituals that use rape to initiate or even punish its members, such as in some gangs. Why does this happen? Why?

The answer is as plain as a single drop of water, yet is as deep as an ocean. The plan is systemic: to turn the broken soul away from He who loves all souls by attacking the body, crushing the soul, wounding the spirit- -kill, steal and destroy.

STEAL*: (the rape itself):*
The violated body is now regarded as unholy and unworthy of goodness.

KILL*: (the very soul using hatred, shame and fear):*
The crushed soul is crippled and unable to function as before. Darkness, contempt and indifference result.

DESTROY*: (the spirit with sorrow, pain and confusion):*
The wounded spirit is left lifeless and struggling to survive. Callousness, agnosticism even bitterness against God.

One interesting thing I must note: Sometimes the attack of the enemy backfires and the victim turns to the Lord immediately for help, comfort and peace. However, that does not appear to be the normal reaction. The whole plan is to take from you what is sacred and turn you into something that you are not. To take a decent man and make him hateful or mean. To take an outgoing woman and make her reclusive and murderous in her heart. Or take a friendly boy and turn him into an effeminate spectacle. Or even a sweet girl and turn her into a hardened whoremonger.

Anything but what God originally intended. Anywhere but where the Lord planned for them to go. Becoming anyone but who God designed them to be. Examine the following metaphorical descriptions, if you will, to consider a few of the deceiving channels the enemy uses to snare someone into becoming a rapist:

1. Opportunist- Just looking for a chance to get away with fulfilling restless, dark imaginations.
2. Instructor- One who seeks to teach the victim a lesson. Either how to dress, where to be, how to behave, how important it is to be serious or take him serious or respectfully.
3. Avenger- The one that seeks to get back what he emotionally lost directly or indirectly because of the victim. Varying from ego, pride, reputation, confidence, self-worth or respect.
4. Game-hunter- The one who hunts the victim to display or prove his manhood or libido. Also, those who feel that their sexual identity is in question by others.
5. Gambler- One who victimizes another because of a pledge, bet or game.
6. Investor- One who victimizes someone that he has invested time, money, gifts or attention to. He expects a return and will not take no for an answer.
7. Rite of Passage Taker- One who is inexperienced in his understanding of sex and wants to prove his maturity and readiness for manhood, by victimizing someone weaker than himself. Young, adolescent and introverts who victimize others fit this area.
8. The Hater- One who directs his anger towards another through a sexual attack. The victim may only remind the attacker of who he really hates.
9. The Insatiable- One whose mental weakness and dark conscience causes him

to want sexual gratification from anyone at anytime he can get away with it. Gender and age are irrelevant.

10. The Predator- One who looks for those who are weak, young, naïve or simply alone to build his trophy case. He may even threaten their lives in order to return to the same victim again.
11. The Challenger- One who wants to humiliate others who are strong, popular, highly attractive or successful, in an attempt to weaken or stop her or him from continuing to show up his seeming inadequacies and shortcomings.
12. The Super-Hero- The one who refuses to accept the victim's rejection because he believes that she really needs him. He will try anything to prove he is the man she needs, even putting her in danger himself and saving her.
13. The Punisher- The one who believes that the victim deserves to be punished for what she has done wrong or for being a 'bad-girl'. He may prey on prostitutes, delinquents, rebellious teens, partygoers and late-night club frequents. He may even be a law enforcer who corrects his victims for breaking the law in his own twisted way.
14. The Love-Starved- One who, out of idleness or jealousy for love and affection, takes what has not been given.
15. The Carrier- One that repeats the offense that he experienced, infecting others out of anger, loss of ideals and hopelessness.
16. The Ruler – One who believes his authoritative position affords him the right to sexual access to whomever he pleases, especially if he feels he is idolized by those he leads.

This only highlights a few common evils I have learned about. Regardless of the general plan that the enemy of man may carry out, he cannot control the outcome. People who trust in the Lord do not have to be subjected to the enemy's constant advances. He has no power over God's people.

It is said that the best defense is a good offense. Though it is obvious in sports, that principle may not be as clear in other arenas. Besides, life is not a game. However, the standard is still true. Being on the offense is when you advance toward victory. We too often stay in a defensive mode, trying to figure out how to protect what we have. Unfortunately, too often our human efforts are futile and we feel crushed when we are not able to control what happens to us. We blame the people or the circumstances for what happened without realizing the dark force that was really behind it. Ignoring the enemy won't make him disappear. We must categorically size up our enemy in order to build up our offensive mode of attack. Let me explain.

In sports, the offense identifies the strongest opponents and prepares to avoid them or elude them, all while focusing on the goal, which is to score against the opponent. In a legal battle, the offensive team studies the evidence that the defense plans to use as proof of innocence and counters it with questions to invalidate and discredit their submission to nullify it. In war, the leaders must consider the potential strength of the enemy based on their known resources as well as their own capabilities to combat them. (*"...Lest Satan should get an advantage of us: for we are not ignorant of his devices..."*

2 Corininthians 2:11)

In all of this, there is a gathering of information first. Knowing the enemy's abilities and tactics is foundational for a plan of victory. Right about now you might be wondering how this analogy connects to a victim of rape? Just like the athlete, the victim must focus on the goal no matter what. The goal is to live and live well, refusing to succumb to the hardships and challenges that arise.

Remember the list of the rapists' mindsets? That should give you an idea of the type of emotional triggers that have commonly set rapists off into such mental depravity enough for you to avoid those who are weak-minded when you recognize them. Furthermore, if you currently dwell with a rapist in an unavoidable place, still you can be victorious. You 'score' on the enemy when you thrive with the zest for life and the quest for destiny that we were all born with. Still going after your dreams, achieving your goals and fulfilling the purpose that God created you to live out in spite of your present but (because of your faith) very temporary situation.

I still glean from the resolve of the African American slaves to keep living and hoping for a better life. They were bound with chains, but not in their spirits. They sang songs to the Lord to encourage themselves, for they knew that no one could enslave a soul. If they had not believed so earnestly, tried so valiantly and persevered, that entire generation would have perished. They learned to read and write and handle business. They helped each other to survive and keep going. They risked their lives to make an easier life for their children's children. They were slaves, but only physically.

"I had fainted, unless I had believed to see the goodness of the Lord in the land of the living. Wait on the Lord: be of good courage, and He shall strengthen thine heart: wait, I say, on the Lord."

Psalms 27:13-14

Right now, in some dark prison cell or isolated room, there are dear souls that are regularly being used for the sexual gratification of some sick person. But they are still alive inside, even if they feel as if they have died. Realize that no rapist can ever imprison a soul; so don't let his actions quiet yours. The enemy has no power over the children of God. He abuses your body, but he cannot make you stop singing, smiling, laughing, dreaming--living. He cannot take your future away. You have to make the decision to keep on because you are not your body nor are you defined by what happens to it.

I saw a news report about a young girl that was abducted by a man in broad daylight. The fascinating thing about the girl is that she kept her wits about her. Instead of giving into the fear she unquestionably felt, she paid close attention to what was going on. While in his car, her vision was impaired because he had forced her into a container, but her mind was sharp and she noted how many times he turned the car and about how long they drove. With that she was able to later identify the vicinity of where she was when she escaped long enough to call for help. She was determined to see the man brought to justice, and he was.

An assault is not the end, even if it has been continual and ongoing. Every rape is not a one- time occurrence, it is commonly known that many suffer from sexual abuse for years. It is therefore imperative that the <u>offensive</u> frame of mind come into action in a very calculated way. The goal is to live and live well, remember? Now, to live under the oppressive appetites of a sexual predator is no way to live. Therefore you must first believe that you can and will be free from it.

Consider this carefully: the more you build up your spirit with God's word, the more your spirit will be strengthened so that you can, in faith, ask God to deliver you. Your faith is key, without it, you won't be able to stand. We have talked about memorizing bible verses throughout this book, but you must now apply them to your life in every situation. Scriptures such as, *"No weapon that is formed against thee shall prosper..."* (Isaiah 54:17) or *"...the Lord is the strength of my life; of whom shall I be afraid?"* (Psalm 27:1b). *"God is our refuge and strength, a very present help in trouble."* (Psalm 46:1). There are thousands of passages in the written word of God that the Lord will accomplish when you believe in Him.

You should begin to speak these words of life out loud to yourself constantly and pray for the Lord to move you out of that situation and away from that sexual evil. Even if- please understand me- even if your prayer is not immediately carried out, you still must build up your faith. Pray the entire time, no matter what is going on. Don't ever stop fighting to keep praying. Pray out loud and if you must, silently, but just keep praying. I can liken prayer to building a fire, it may not catch right away, but you have to keep at it.

"I can do all things through Christ which strengtheneth me."
Philippians 4:13

Memorizing and speaking out the scriptures as well as praying are key offensives, but you must also encourage yourself with music that builds your soul. Christian music is written in many styles, find what you enjoy and what stirs you. Music flows inside of you like water and is easily digested by your soul, remember that. Before you know it, you will have your own song inside of your heart. Begin to sing these songs to the Lord, believing that He is listening, because when you sing believing that He hears you, He does.

"...speaking to yourselves in psalms and hymns and spiritual songs, singing and making melody in your heart to the Lord..." *Ephesians 5:19*

One of the most important practices you should develop is to become a worshipper. Praise the Lord for life. Praise Him for the bright future that you have. I know it that may not be the easiest thing to do, especially if don't feel like it, but trust me, the more you focus on the Lord, the more your life will change. *"Draw nigh to God and He will draw nigh to you."* *James 4:8*

Finally, you must connect yourself to other believers. Even if you cannot get out to church, God will provide someone to stand with you because you need it. It may be a relative, neighbor or even a minister on the radio, just get yourself connected to someone else who can pray with you and believe for your deliverance. This fight is going to take all the reinforcements you can get. Remember, you are not fighting against a rapist, but a spirit that is trying to keep you from reaching your destiny by using sexual slavery.

The good news is, God will deliver you if you believe. He can even deliver you if you aren't sure what you believe, because He will move on the prayers of others that are praying for you. So don't think you have to have all the right words or pray a certain way or know a lot of Bible verses. You need to use the faith you already have. The faith that made you read this book, believing for an answer, is the same kind of faith that will make miracles happen. *"But without faith it is impossible to please him: for he that cometh to God must believe that he is, and that he is a rewarder of them that diligently seek Him."* *Hebrews 11:6*

It doesn't matter if you have a life sentence in a prison or live isolated in some forgotten land, the Lord can make you live out your time in peace. You may even be married to a man that violently takes advantage of you sexually, (which has nothing to do with marriage), God can even change him. Or for those dear ones who are waiting for a chance to run away, God can make that spirit leave you instead and never return after you again.

For those who are bound by a (so-called) religious community's practices of sex, please understand that God is holy, loving and good. The Lord would never violate or defile you for any reason, nor would He require or condone such behavior from anyone else. When King David acted as if he had a right, as king, to take Bathsheba, the Lord God punished him and blessed her, remember?

The Lord is gentle and kind. He would never direct anyone to offend a child. In fact, He warns us sharply against it. (*"But whoso shall offend one of these little ones which believe in me, it were better for him that a millstone were hanged about his neck, and that he were drowned in the depth of the sea." Matthew 18:6)*

Sexual crime is one of the deepest forms of offense that is committed. It leaves an indescribable and indelible mark on the soul that causes an abrupt and destructive cycle of change. As strong as it may be, however, it still cannot defeat the will of God. There is still, in each of us, the ability to choose the Lord's way, in spite of the bad things we experience. His way always produces life.

"...I have set before you life and death, blessing and cursing: therefore choose life, that both thou and thy seed may live: That thou mayest love the Lord thy God, and that thou mayest obey his voice, and that thou mayest cleave unto him; for he is thy life..."

Deuteronomy 30:19-20

Determine in yourself that you will choose life and not death. Keep on breathing and striving for that true destiny in spite of what you are currently enduring right now. You must make up your mind that you will do all that you can to please God above all else. You can't be responsible for anyone else's actions but you must rule over your own. Stay true to what is right, leave the outcome up to God. When it is all over, and one day soon it will be, the Lord will reward you for your faithfulness to Him. Your faith will always be more powerful and much more effective than anything the enemy of your soul can ever do. Inside, you are always free.

"Ye are of God, little children, and have overcome them: because greater is He that is in you, than He that is in the world."

I John 4:4

Journal Entry About God's Liberty:

Family & Friends

Much of this section was written with family in mind. It is critical that you understand the victim's state of mind.

They feel trapped in a place that is isolated, dark and dangerous. Just getting up and getting dressed is a major accomplishment for many of them. No victim needs to be given the 'silent treatment' or the 'cold shoulder'. Even if they didn't heed your warnings of "Don't go out with him..." or "Why do you dress like that?" This is not the time to say "I told you so." Indifference is a very hurtful way to communicate.

You simply cannot blame the victim. They feel bad enough. Don't kick them while they are down. Never, ever question them- -ever.

You are a key person in the victim's life. Your prayers, your words of life and your conduct towards and around them can make a world of difference.

The attacker left a trace of shame on the victim that only real love can melt away. That's where you come in. You must treat the victim as if they are already whole, because your faith is stronger than the present circumstance.

The victim is not pitiful so don't pity her. Pray! Find encouraging books, audio-recorded sermons, and Christian music. All of these will help with the recovery process.

Help your loved move away from being defined as a victim. The enemy has no power over who he or she is- - let them know that.

At the end of this book, I have included a prayer similar to what I prayed when I'd been stalked as a young woman. It wasn't until then that I realized that I was neither a victim nor a target. After years of suffering before that time, I then finally saw the end of that battle.

No one has power over me, none. For I am in Christ Jesus and He has overcome the world.

section four:

"...and recovering of sight to the blind..."

"Now Jephthah the Gileadite was a mighty man of valour, and he was the son of an harlot: and Gilead begat Jephthah. And Gilead's wife bare him sons; and his wife's sons grew up, and they thrust out Jephthah, and said unto him, Thou shalt not inherit in our father's house; for thou art the son of a strange woman. Then Jephthah fled from his brethren, and dwelt in the land of Tob: and there were gathered vain men to Jephthah, and went out with him.

And it came to pass in process of time, that the children of Ammon made war against Israel. And it was so, that when the children of Ammon made war against Israel, the elders of Gilead went to fetch Jephthah out of the land of Tob: And they said unto Jephthah, Come, and be our captain, that we may fight with the children of Ammon. And Jephthah said unto the elders of Gilead, Did not ye hate me, and expel me out of my father's house? And why are ye come unto me now when ye are in distress? And the elders of Gilead said unto Jephthah, Therefore we turn again to thee now, that thou mayest go with us, and fight against the children of Ammon, and be our head over all the inhabitants of Gilead.

And Jephthah said unto the elders of Gilead, If ye bring me home again to fight against the children of Ammon, and the Lord deliver them before me, shall I be your head? And the elders of Gilead said unto Jephthah, The Lord be witness between us, if we do not so according thy words.

Then Jephthah went with the elders of Gilead, and the people made him head and captain over them: and Jephthah uttered all his words before the Lord in Mizpeh." Judges 11:1-11

Chapter Ten: Mistaken Identity

We have no clear idea of who our children will become. Just because the manner surrounding the conception of the child may have been undesirable does not negate the worth or value of the child's life.

Jephthah was put out of his home by his own family and the people in his community because his mother was a harlot. They refused to let him have his inheritance, that is, until they realized how much they needed him.

The great thing about Jephthah was that he never forgot who he was. He chose to be a man of valor, diplomacy and strength. More importantly, as scripture later tells us, he was a man of conviction and prayer. Every child has a destiny. No one should be so presumptuous as to think otherwise.

The unfortunate thing is that not every child who experiences great pain is as determined and ambitious as Jephthah was. Although he was not victimized by sexual crime, he still had much to contend with by being so publicly demoralized. He did not have the support that young men usually need to grow up and mature well. When he should have received love, he instead was harshly rejected and mistreated by his own family and community. Had Jephthah not been a born leader, he may not have been able to move on with his life.

Even with what Jephthah endured, as emasculating as it was, he held on to his faith and sense of purpose. That is why I believe he is such a great example for those who have been knocked down emotionally. His situation may be different, but pain is pain. Jephthah overcame his pain and was able to accomplish what he was destined to do. If he can do it, you can do it.

Sexual assault is hard to come back from, especially for a child. When an adult is raped, he or she, in a sense, becomes a child. When children are raped, they stay childish emotionally many more times than not. After an assault, each individual usually behaves very differently. The normal behavior or way of thinking is altered in immeasurable ways that can range from anxiety and shyness to exhibitionism. There are also often expressions such as rage, cruelty, anorexia, obesity and depression that can go unchecked.

More extreme are the identity issues including effeminism, homosexuality, voyeurism and prostitution. The crisis is not primarily the assault that occurred, but the struggle that follows, taking years, many times, to overcome. That is why protecting our children, even if it angers or frustrates them, is so important.

Jephthah was mistreated by his own half-brothers, even though they knew he was a man of valor and strength. When it came to dividing the inheritance, all they chose to see at the time was his pedigree or lack thereof. But they were mistaken. His true identity had nothing to do with who his parents were. This mistake of identity doesn't just happen by those who look at others the wrong way, many times, we look at ourselves blindly.

Often times, after being assaulted, people tend to look at who they are based on how they feel instead of still seeing themselves as humans created in the very image of God. If the assault left us feeling confused, we do things that are completely contrary to our personality. If we are left feeling powerless, we overdo our expression of strength. When we are afraid to be what we were before the assault, we blindly masquerade ourselves as someone else.

It is this inner blindness that even makes us behave in ways unlike what God desires. I will identify some of the less obvious areas of this 'blindness', to bring some clarity as to the dangers they can cause.

Pornography

I have learned that pornography is a trap into an isolated world that stunts social civility and compromises ones ability to show respect to others. It is a disgraceful depiction of the lewdest type of human behavior that eats away at the conscience like rust does to metal. Rust is a type of fire; though it does not feel hot and does not generate smoke or completely burn immediately, it still burns and over time is just as entirely destructive as any other unchecked flame.

Pornography may not have all the obvious ills apparent, but it is one of those acidic behaviors that is capable of destroying a person from within. Only by considering the will of the Lord can one begin to escape from this dangerous, unclean practice. Again and again we hear the findings of investigative reports of rapists, sadists, pedophiles and murderers that seem to all have indulged in pornography. I refuse to believe that the connection is coincidental. Pornography is a secret sin and, continually engaging in any sin changes you. Your conscience won't even be sensitive enough any longer to convincc you that it is wrong.

"Now the Spirit speaketh expressly, that in the latter times some shall depart from the faith, giving heed to seducing spirits, and doctrines of devils: speaking lies in hypocrisy; having their conscience seared with a hot iron..." *I Timothy 4:1-2*

Voyeurism, Masturbation and Exhibitionism

The reason why voyeurism, masturbation and exhibitionism are so powerful is because each is a version of self-worship. If you are one who practices this, you become an unpaid porn-actor in a sordid show produced by a demon and starring you. Why? To devalue you, disgrace you, discredit, disrespect and displace you from God's revealed glory and presence. The devil wants to keep you away from God by engaging you in secret sin.

"Behold, the Lord's hand is not shortened, that it cannot save; neither his ear heavy that it cannot hear: but your iniquities have separated between you and your God, and your sins have hid his face from you, that he will not hear. For your hands are

defiled with blood, and your fingers with iniquity; your lips have spoken lies, your tongue hath muttered perverseness."

Isaiah 59:1-3

Satan is working his craft of destroying any natural desires you have by ruining your conscience and stealing your understanding of who you are and why you were created. Your vision of why God made you is replaced by the devil's perverted offer of a good time. There is nothing good about sin. Natural, yes, but good- never. If we must choose between what is natural and what is good – choose the good.

For one thing, voyeurism, or the practice of obtaining sexual gratification from seeing sex organs and sordid, scandalous sexual acts, is a deep trap. It is insatiable and will not allow you to merely practice secretly. It builds up the imagination and desire until an actual physical experience takes place.

It is common knowledge that humans naturally process what they see by utilizing our other senses to get a full experience. This is seen even in infants who will at length attempt to touch or even taste what they see, in order to fully experience it. Eve was tricked by the serpent into examining the forbidden fruit and she ("...saw that the tree was good for food and that it was pleasant to the eyes, and a tree to be desired to make one wise, she took of the fruit thereof, and did eat...(Genesis 3:6); thus, her demise, and ours began. That old serpent is still tricking the children of Adam and Eve in the same way to this day, to see, touch and experience what is forbidden. As ever, he never discloses the true consequences of the action.

Allowing yourself to lust after others by gazing, leering and peering for sexual stimulation only gives you an artificially induced reaction that will leave you craving more. The more you see, the more you want to see until it no longer matters *who* you see or how you voyeur to quench your appetite. What's to stop you from degenerating from pornography to child pornography? From photos to private windows? The same driving force that is leading you to your private, delusional ecstasy won't stop until you either resist it, or end up in a private cell or worse, a mental breakdown. Anything to keep you from following after God.

It is understandable that when one has been sexually defiled, it is a challenge to diminish the confusion that has resulted. Attempting to prove or find one's sexual identity is the excuse many have adapted for getting involved in voyeurism. It may seem like a ready answer for peer pressure, but it is not the answer to the real crisis within. Voyeurism is a flood that is not easily dried out. Leave it alone. It may start out as a way to prove your sexuality, but it won't lead you into a true love relationship because it breaks down your sense of value and respect for others.

A man that practices such won't be able to connect with a good woman because he will no longer have the conscience to show her the respect that she will require. On the other hand, a good man will not want a woman who behaves herself in promiscuous ways. Don't let pornography and sexual licentiousness turn you into someone indecent with whom no one can abide. Find your identity in the plan of God for your life.

Masturbation is idolatrous sexual-self-worship. Through this the enemy has unhindered access to your body without driving someone to force it from you, date it from you or deceive it away from you. If he can succeed at getting you to sin against your own body, and get you to behave in lewd, vulgar, repulsive and unclean ways in the sight of God, he has taken one more worshipper out of his purpose and out of God's will.

Taken them away from the place where their greatest dreams, visions and potential are reached. Taken them away from what is wise and wholesome and good. Deceiving them into being unholy, to ensure that they will not be pleasing to the all holy God. It is commonly stated that self -sexual practices are "only natural"; but it is so much more than that- - it's supernatural. There is more going on than mere self-gratification. It's about taking your eyes off of the Lord, which, even for a moment, is dangerous.

No one can worship God and their own flesh at the same time. Because the practice of masturbation forces you to use your imagination in order to be fulfilled, you literally have to shut off the voice of your conscience every time you indulge in it. The more you do so, the more likely you are to regard sin as an acceptable way of life.

It is not unusual to find one that has been sexually offended, involved in some dark, sexual practice. It is even understandable, but it does not have to end up that way. There is still a choice. Don't allow the crime committed against you cause you not to care about your own soul so much that you give up your quest to walk with God.

"Follow peace with all men, and holiness, without which no man shall see the Lord: looking diligently lest any man fail of the grace of God; lest any root of bitterness springing up trouble you, and thereby many be defiled; lest there be any fornicator, or profane person, as Esau, who for one morsel of meat sold his birthright. For ye know how that afterward, when he would have inherited the blessing, he was rejected: for he found no place of repentance, through he sought it carefully with tears." Hebrews 12:14-17

The plan of the enemy is to make you fall away from the purpose of God for your life. You have to have a genuine respect for God's decision to create you by striving to walk in your divine purpose. When you allow yourself to lose sight of who you really are, you can find yourself doing anything, even repulsive and demeaning things. Bitterness can bring you to that place where you no longer care to respect yourself. This can show up in various ways, such as exhibitionism.

Exhibitionism

Exhibitionism is the child of the devil's first inception of the "Hey-look at me;" "Don't you want me?" "Don't you wish you could be with me?" attitude. All it is, is pride in you own flesh, which is also the very thing that got Satan kicked out of the heavens. Don't let him get you ousted too.

This is one area that plagued me for a long time after my experiences with sexual crime. I wanted to feel some level of control over feeling so vulnerable. It may not have blatantly shown up to the world around me, but my imaginations were strong in this area. I wanted to be the one to say when I would engage in sex, not to have it forcibly taken from me. There would be many times when I would envision myself as this free soul that no one could touch unless I wanted them to, no matter what I did. My sensuality became my own subtle, personal power.

To be in a setting where I knew I couldn't be hurt, say a public or well-protected place, and, (with my promiscuous choice of clothing), almost taunt those that had hurt me before, was my silent revenge. But it was still just sin and I really had no excuse to behave or imagine myself in that way. Even worse, I put the men in my former relationships in very compromising positions and if the Lord had not intervened, I would have caused them to sin as well.

That behavior was not right no matter what I was trying to prove, what I was trying to forget or how much I wanted to make myself feel better. I belong to the Lord. My body is not to be displayed to taunt, intimidate or fascinate anyone, nor is it to be given away as a reward for those who manage to somehow please me. It is the temple of the Holy Spirit of God and is only to be shared in a worshipful and sacred manner with my spouse.

In my early adulthood, I stopped dating completely just to avoid emotional diversion and focus on my destiny. Before then, dating was so distracting, (due to all the hurt in my life that I had not yet been healed of), I only progressively got more away from the plan of

God with each relationship. I compromised my morals and ignored my conscience with each opportunity (relationship) to forget my pain.

If we can be honest, we would admit that dating can really get us off course and position us to make decisions that prove to be regrettable and even sinful. When it is time to marry, and if we are completely submitted to the Lord, He will direct us in a way that will be right in every aspect. Dating at any other time but His is an unwise waste of emotion, time and energy that we will never get back.

The truth is, your body belongs to God, whether you honor Him with it or not. He made it, fashioned it, designed it and has every right to it. Every cell is marked with purpose. Every atom has cause. The thief can only successfully offend God with your body if you allow him to. Do you want to be numbered with those who follow Satan's influence? Of course not.

There is only one way to resist this dark plan of our enemy: Offer your body as a living sacrifice, holy and acceptable to God. (See Romans 12:1) Your body is meant to be the temple of the Holy Ghost, not a shrine of sensuality. Keep your body under control. Possess your vessel – don't obsess over it. Jesus Christ has given us power over sin. In other words, we have the power to behave ourselves!

"Furthermore then we beseech you, brethren, and exhort you by the Lord Jesus, that as ye have received of us how ye ought to walk and to please God, so ye would abound more and

more. For ye know what commandments we gave you by the Lord Jesus. For this is the will of God, even your sanctification, that ye should abstain from fornication: That every one of you should know how to possess his vessel in sanctification and honor; not in the lust of concupiscence, even as the Gentiles which know not God: That no man go beyond and defraud his brother in any matter: because that the Lord is the avenger of all such, as we also have forewarned you and testified. For God hath not called us unto uncleanness, but unto holiness."

I Thessalonians 4:1-7

Your normal sexual passions are just indicators that everything is in working order as it should be. Focus on what God intended for your body even when you are distracted. For example, when passionate dreams arise, cast down those imaginations by meditating on scriptures about holiness. *"Having therefore these promises, dearly beloved, let us cleanse ourselves from all filthiness of the flesh and spirit, perfecting holiness in the fear of God." (2 Corinthians 7:1)* Take captive every thought immediately that goes further than the will of the Lord would have you to go. Don't allow yourself to entertain any thought or activity, (i.e., petting, sexual outer-course, etc.) that will dishonor the Lord our God. *"Submit yourselves therefore to God. Resist the devil, and he will flee from you."*

James 4:7

Even if you feel that behaving sensually gives you power (via making those who desire you weak), you have to let that feeling go. We are to worship the Lord and Him only are we to serve. We aren't to make ourselves so desirable in our appearance or mannerism that we are distracting others wherever we go. If too many heads are turning towards or away from you, if there is too much gazing, if comments from young, old, male, female and even children are directed at your attire in a negative way-you have probably gone too far and you need to change your style of dress.

It is the Lord's attention you need to hold, no one else's. Putting the private areas of your body on display is disrespectful and inconsiderate, regardless of how becoming your body may be. This behavior also robs people of the chance to know you past your uncivilized presentation of your flesh. Don't force people to react to you because of your indecency.

God is not mocked, you will reap what you sow. Therefore, when you are tempted to sow to the wind, that is, just thoughtlessly give in to sin, focus on the Word of God and speak it to yourself. Seek God's deliverance for any and every struggle. His power is attainable and we are expected to avail ourselves to it. There is no excuse to stay in sin once we have come to Christ. Being single, in a contentious marriage that is bereft of intimacy or simply being lonely is no reason to defile yourself and become abominable before the Lord God. Here are a few practical ways of addressing this useless behavior <u>after you have repented</u>:

- Change your wardrobe including overly sensual scents -start with the trash can...(this includes men as well). Pheromones are natural chemicals that your body produces to prepare your spouse for physical oneness with you. You do not need to use products that over-intensify this natural occurrence. Be clean, groomed and fresh, not enticing.
- Guard what you see, hear and read. Sensual programs, movies, plays, music, romance novels and even magazines with articles about sexuality are dangerous and leading.
- Be mindful of the company that you keep. Their behavior, good or bad, will effect you sooner or later. Therefore, surround yourself with those who love the Lord and live holy before Him without guile.
- Practice being friendly without being flirtatious. Learn the difference and mark anyone who confuses your kindness by demanding respect.
- Memorize the following scriptures: Galatians 5:1; Hosea 10:12; I Timothy 6:11;Titus 3:8; Luke 9:25; I Corinthians 9:27.
- Stay rested, eat properly and get good exercise. (Taking care of your body helps to prevent fatigue and hormonal imbalances that make resistance to sinful practices more difficult.)
- Memorize any particular verse the Lord shows you personally.
- Stay prayerful and fast frequently.

When the Lord has placed a high calling on your life... when He wants to exalt you...When God himself desires you to be in the forefront to affect the masses – the enemy of your soul will attempt to disrupt that plan with opposition. Why? To make you miss your call, be brought to the lowest place possible, put you way in the background and far away from the very masses you are supposed to help.

You were created on purpose with a purpose for the good of all. If you are caught up in a deceitful web of secret sin, it tangles up your potential. Guilt and darkness of soul clouds your ability to believe the best about yourself. Without even realizing it, you condemn yourself to being less than who you are, by wasting time trying to get over the pain of the gulf you feel from your own sin.

Try to remember that you are more than what you feel. Even if you don't see it or imagine it, you must press yourself to that place of destiny. Don't allow anyone or anything to stand in the way of you becoming the person you were born to become.

"...but this one thing I do, forgetting those things which are behind, and reaching forth unto those things which are before, I press toward the mark for the prize of the high calling of God in Christ Jesus." *Philippians 3:13-14*

Remember Jephthah. He didn't allow the past, the people or the circumstances stop him from being a man of integrity. He sought the Lord for help on how to accomplish what he was called to do. *"See then that ye walk circumspectly, not as fools, but as wise, redeeming the time, because the days are evil. Wherefore be ye not unwise, but understanding what the will of the Lord is. And be not drunk with wine, wherein is excess; but be filled with the Spirit:..."* *Ephesians 5:15-18*

The opposition couldn't crush Jephthah because he kept his head up and looked to the Lord. They thought they could determine his destiny but they were wrong. They mistakenly identified him as a weak man, but he wasn't. When he could have repeated family history and engaged in shameful sin, Jephthah chose instead to live a life of respectful influence, leading even those that others rejected, and wait for the Lord to change his situation. If he could do that by relying only on God, I am quite certain that we can do the same. Don't allow yourself to be identified as less than who you were created by God to be and don't behave yourself as if you are not who God says you are. Jesus already overcame the world; embrace the power of the freedom He provided.

Chapter Eleven: What Child Is This?

Such a wonderful Christmas carol we used to sing. Part of the song plays: "What child is this who laid to rest on Mary's lap is sleeping…?" The question in itself makes you want to wonder at who the child is that the song is about and how precious he must be for such a sweet melody to be ascribed to him.

That's why I named this chapter the way I did. Many times when a child is conceived as a result of a sexual assault, the child is unwanted. The pregnancy is frowned upon and the mother is pitied to shame. Some women are even scorned. The entire situation from the rape itself is bad enough, but it is compounded with complexity when conception occurs.

What is a woman to do in this situation? Sorrow mounts and depression pulls like a hidden undercurrent. The walls seem to close in and there is seemingly no easy answer. To just move past the attack is not as likely when there is a constant reminder of what happened growing and moving inside of her. The range of emotion is hard to empathize, as each woman's situation is different.

Some may have the ability to hide themselves for a while. Others can relocate or temporarily leave the area. But many have no place to go and bear the shame of what took place with every turning month.

Her soul aches as she yearns to love this tiny life she carries, yet separate herself from all traces of the attacker.

If this were not hard enough, just imagine the torment of the woman who had longed for and waited to have children only to find herself in the condition that she wanted, but in such a hurtful way. This is not what she hoped for or anticipated. The mental anguish is mind numbing. It is enough to make her consider aborting the child, (yes, even the very ones who stood against abortion before they were attacked may find themselves in this emotional fight). It is part of the woman's survival instinct to want to cleanse herself from the horror of what happened to her. Her hatred for the rapist may grow as steadily as does the child, causing the reluctant mother to despise herself for feeling so badly about an innocent baby.

The very thought of having to look at and care for a child that resulted from the rape haunts her almost as much as the assault-if not more at times. "What if the child looks like him…or worse, acts like him when he is older?" "Will I be able to love this child?" Such questions race through the woman's restless mind, disrupting the healing process.

Every now and then she may be able to escape the thought of the misery of the conception and enjoy her experience, but those moments are too few, I'm sure.

How do I know so much about all of this? The Lord is all-wise and He answers the questions our hearts express. (Hebrews 11:6). He answered my questions even about areas of this issue that I did not experience, because I wanted to know what He had to say. "…he that cometh to God must believe that He is, and that He is a rewarder of them that diligently seek Him." My experience was not even close to that of others, yet pain is pain and only God knows what we feel.

Our feelings can keep us from seeking the Lord whole-heartedly and we cannot let that happen, especially when a new life is at stake. First, understand that those feelings are just that. Feelings are about as stable as clouds, you can't control them and you certainly can't count on them staying very long. Even though these feelings come and go, they can be just as menacing and devastating as a storm cloud. All the thunder and lightning and torrent of rain is all to be expected in a storm.

Being attacked sexually is somewhat similar to a violent storm. A dangerous tidal wave is caused by a strong earthquake. Likewise, the overwhelming tidal wave of emotion that you are suffering was caused by the sexual sin. When Jesus walked the earth, He commanded violent storms to cease, and the earth obeyed Him by coming to a complete calm. Jesus can calm the emotional storm that you have been enduring just the same. No one has more power than God does. No power is greater. As certainly as that rapist hurt you, God much more certainly has the power to completely heal you, and He will, if you believe.

Prayer is one of the greatest things you can do as you come out of this dark place. Pray about everything. Tell the Lord exactly how you feel. Let him help you. He will comfort you, dispel your fear, strengthen you, give you courage and if that were not enough, He will even work the entire situation out for the good.

If God can make something out of nothing, He can make things that are weighing you down in your mind into nothing at all. Once you begin to rely on Him to help you deal with the emotional toll that all of this has taken, you can begin to trust Him to change your heart towards the baby. What child is this anyway?

Realize, first of all, that God is the master of conception. No man has the power to create life, he only participates in the process. God determines which souls will exist and to what end they will affect the world once they are here, however long or short their span of life. Knowing that God purposely decided that at that particular moment in time, when the enemy aimed to destroy you and further enslave the man who harmed you, God who is sovereign, made good come out of it, and allowed conception. So the child was given the sacred breath of life with purpose, even if you did not participate in his beginning, on purpose.

The second thing to know about this child, is that he or she is already special. God had already put the seed for this baby in you years before you thought of motherhood. The enemy might have noticed how special you are and wanted to make you experience a devastation that would make you forget who you were. Satan will stop at nothing to keep you from accomplishing what God put you here to do. *"For we wrestle not against flesh and blood, but against principalities, against powers, against <u>the rulers of the darkness of this world</u>, against wickedness in high places." (Ephesians 6: 12)* Rape is one of his dirtiest tools. Sickness, poverty, bigotry—all are strong weapons, but sexual assault?, That one he uses to try to halt the great ones he is aware of. But he is not successful. Jesus conquered Satan and all his evil plans. See, this child is of no consequence, but of purpose. The enemy was after you to make you turn from God and live opposed to your own destiny. Somehow, that precious baby is connected to your purpose.

That leads me to the third truth about this child. Could it be, that the baby is purposed to accomplish a great feat in life here on the earth? Could it be that this fact was spoken over your family, generations before you were born, as was true of many in the bible like John, Moses, Josiah and of course Jesus Christ himself? Was the enemy's plan to stop the baby from coming, or worse, press *you to hate the child so much that the child grows up offended beyond his ability to overcome?*

Why did Jesus speak such harsh words about offending a child? Because you don't know how the child's life is predestined to be. You don't want to be the one held responsible for the missed calling of the one who would have cured cancer or AIDS. Or the next great motivator of millions or even a preacher that guides countless souls into the kingdom of God. Even a specialist that saves lives by bravely diffusing bombs. Maybe this whole assault, wasn't really an attack on you so much as it was one on this particular baby that you are destined to raise.

With that in mind, how do you see the child now? What child is this? Who will he become? What might she do? When will you know? The answers to these questions will only come if you do the good you know to do. Just love that baby. Keep him or her safe and well. Nurture and treat him or her with care and grace. Don't say 'I can't do it, but rather, 'I can't do it without You, Lord'. This truth just happens to be a simple prayer. Prayer is all I am really asking you to do. To raise this child the right way, prayer is the thing you must do continually…for God will, quite certainly, help you to do the rest.

Trust Him. Besides, God, who is the Father of life, is the baby's real Father. Let Him father your baby. Teach the child His words and His ways and let the Lord show you what to do. The church can teach you how.

When situations get hard or when the child does things you don't understand, ask the Lord to take care of it, there is nothing too hard for God and there is nothing impossible for those who believe.

Of course there may be a resemblance in the child that triggers your memory of what happened, but you can't dwell on it. Jesus is the Prince of Life. Enjoy your motherhood. Enjoy your child, whether she is your first or in the middle of four others. A child is always a blessing. Moreover, God giving you the ability to bring forth life, is a blessing in and of itself.

Of course, I cannot fail to encourage you to recognize that your child will forever be precious to God. He will take care of the baby, protect and nourish and anything else that you need for her. You must try to realize that you are not alone in this. The Lord will give you wisdom for handling every situation that arises concerning this child. Try to stay focused on the goal. Don't allow yourself to be distracted into recalling the darkness of the conception, live in the day that you are in. Enjoy being a mother. Watch the child grow in peace. This is my prayer for you.

There is another account of scripture that is well worth mentioning here. It is the story of Lot and can be found in Genesis 19:29-38. Rape is horrible in and of itself, but when a family member is the one that has committed the crime, the pain is far worse. Lot's background is a little bizarre because he was the one that was sexually taken advantage of by his own daughters after they made him drunk. They did it to get pregnant, believing that they were the only ones left in the world after seeing the cities where they lived destroyed as punishment by God because of the people's rampant sin. The scripture tells us that God sent angels to get Lot out of the vile cities of Sodom and Gomorrah before they were destroyed. The condition that those cities were in is not the focus of our study right now. We want to look at what happened with Lot and his daughters.

According to the text, the girls took it upon themselves to continue the bloodline by getting their father drunk and getting pregnant by him. Apparently, they didn't' realize that the same God that saved their lives would also provide for the continuance of their posterity.

Why and how they did it is recorded. How Lot felt afterward is not known. Actually, I never found any quotes from him after that. What went through Lot's mind? He was defiled sexually by both of his children. That had to be mortifying and heartbreaking at the same time. Many never speak of the dark kinds of things that go on. It is a shameful fact that within families, behind closed doors and in secret places, unthinkable sexual acts are taking place. It must have been hard for Lot to know that his sons were also his grandsons.

And to think, usually when we regard sexual crime we think of acts that men do to women, yet the first sexual crime that actually occurred in scripture was an incestuous act on a man by his two daughters.

We cannot say that it was Lot's influence that caused the girls' behavior because he was a righteous man in God's eyes. Look at this passage:

"And turning the cities of Sodom and Gomorrah into ashes condemned them with an overthrow, making them an ensample unto those that after should live ungodly; and delivered just Lot, vexed with the filthy conversation of the wicked: (for that righteous man dwelling among them, in seeing and hearing, vexed his righteous soul from day to day with their unlawful deeds; The Lord knoweth how to deliver the godly out of temptations, and to reserve the unjust unto the day of judgment to be punished..." *2 Peter 2:6-9*

From our story it is very clear that Lot had to contend with much. We can also gather from all of this that his daughters most likely did this out of either ignorance or due to the evil influences of where they grew up.

At any rate, what happened, happened. Lot was still a righteous man before God. The Lord was not going to punish him and his innocent sons/grandsons from what his daughters did to him.

No matter what transpired, no matter the disgrace or the double shame of being violated by his own family, God still honored him. The Lord never forgot Lot or what he stood for. He blessed him. Although there was quite a bit of trouble in the family for a while, God faithfully remembered His promise to Lot's Uncle Abraham concerning the righteous. The oldest son of Lot's seemingly unfortunate union, became the father to a nation of people called the Moabites.

From that family many generations later, a woman by the name of Ruth was blessed to become the great-grandmother of King David. As you may know, from the lineage of King David, came the Savior of the world, Jesus.

See, this confirms for all of us today that every child is important and every life matters, no matter the manner of conception. God can work out everything and produce goodness in spite of it all. Lot's blessing as a righteous man was not cut off because of his family's sin. Your situation is really no different. If you have dealt with the pain of sexual sin in your family, don't lose heart, God can still work everything out. He is the one who can make it all better. Keep hoping. Keep believing. *"Is any thing too hard for the Lord?..."* (Genesis 18:14a) He can work out any problem, just watch and see.

There is one last area that will help you, even if the child you bore is now an adult. Speak blessings over that child. Read God's word to him regularly and be specific when it comes to promises that will stir his faith. The child needs to know who his true Father is...who our Father is. Here are a few scriptures I believe will be great deposits into the bank of your child's heart, ready for him to make a withdrawal whenever he needs to:

"...I am fearfully and wonderfully made..." *(Psalm 139:14)*

"How precious also are thy thoughts unto me, O God! How great is the sum of them! If I should count them, they are more in number than the sand..." *(Psalm 139:17-18)*

"When I consider thy heavens, the work of thy fingers, the moon and the stars, which thou hast ordained; What is man that thou are mindful of him? And the son of man, that thou visitest him?"

(Psalm 8:3-4)

"...I have loved thee with an everlasting love..."

(Jeremiah 31:3)

"...thou art mine..." *(Isaiah 43:1)*

"...My Father, which gave them me, is greater than all; and no man is able to pluck them out of my Father's hand..."

(John 10: 29)

"We love him because he first loved us." *(I John 4:19)*

"When my father and my mother forsake me, then the Lord will take me up." *(Psalm 27:10)*

"Many are the afflictions of the righteous: but the Lord delivereth him out of them all." *(Psalm 34:19)*

"I will never leave thee nor forsake thee,.."

(Hebrews 13:5b)

"Wait on the Lord: be of good courage, and he shall strengthen thine heart..." *(Psalm 27:14)*

There is another man in scripture that I wanted to introduce to you. He was a man that we don't know a lot about, but what is recorded is so significant, we should take a closer look.

"And Jabez was more honorable than his brethren: and his mother called his name Jabez, saying, Because I bare him with sorrow. And Jabez called on the God of Israel, saying, Oh that thou wouldest bless me indeed, and enlarge my coast, and that thine hand might be with me, and that thou wouldest keep me from evil, that it may not grieve me! And God granted him that which he requested..." *I Chronicles 4:9-10*

Jabez was born in a time of sorrow and his name reflected his mother's grief (it means sorrow maker). But he pleaded to God to enlarge him and make him better than what he was apparently doomed to become, and God blessed him.

Isn't it interesting to note that Jabez prayed that he would be kept from evil? Why was he expecting it? What sorrow did his mother bear him in? Why didn't she name the others names based on her feelings if it was only physical pain? In the spirit of this study, my curiosity prompts me to wonder, did she conceive him as a result of being raped? Only the Lord knows, the specifics are not evident, but it certainly is interesting that her emotionally charged actions were recorded.

What the Bible does show is that a child does not have to be what he is expected to become. Jabez, regardless of what others may have expected him to be, was more honorable than all his brothers. He also had faith, if he didn't, God would never have answered his prayer for a blessing. God is able and willing to change the life of anyone who will just ask in faith, always remember that. Whatever manner Jabez was conceived, what he became is all that ultimately matters. So it is with your baby.

A child is a child and a life is a life. God is the master of conception. Two healthy people can consummately come together a million times and never conceive a child. God can create life inside of a woman whose womb has been dead and produce seed from a man whose loins are aged past use. He even caused a virgin to conceive without the seed of a man. Again and again, the Bible tells us of instances where we learn that when God wants to create a new life, He does it. The enemy is the one who spoils the joy of the plan by aiming to cause the conception process to at times be sorrowful. This is all relative to the original curse. The enemy hates women; it is part of the curse.

If the natural father is a rapist, what does that make his child? A child! If the natural father is a doctor, does that make the child a doctor? No. Specific acts, whether good or evil are up to each individual. Though it can be learned, behavior is not inherited. A child is a child. Just because he or she was conceived in sorrow does not mean that there is no worth or value to him or her. They are not hopeless, nameless, without inheritance and certainly not without purpose. Jabez proved that quite effectively. Please don't allow fear to keep you from loving that dear soul.

To have to carry a child for almost a year and even possibly risk your health to give birth is a lot to handle for any woman. Having to do so in sorrow can make the entire pregnancy dreadful, if you choose to be inconsolable. But there is so much help and support available now, there is no reason to go through this alone.

So open your heart. Allow yourself to love the baby that the Lord has blessed you with. Don't worry about how you will raise him or what he will become. Expect the best and believe for even more than you can imagine. And always remember, the baby did nothing wrong, don't treat him as if he did. You have yet to see who this child will become, just trust God to help you, He is a father to the fatherless.

"*And Abraham rose up early in the morning, and took bread, and a bottle of water, and gave it unto Hagar, putting it on her shoulder, and the child, and sent her away: and she departed, and wandered in the wilderness of Beersheba. And the water was spent in the bottle, and she cast the child under one of the shrubs. And she went, and sat her down over against him a good way off, as it were a bowshot: for she said, Let me not see the death of the child. And she sat over against him, and lift up her voice, and wept. And God heard the voice of the lad; and the Angel of God called to Hagar out of heaven and said unto her, What aileth thee, Hagar? Fear not; for God hath heard <u>the voice of the lad</u> where he is. Arise, lift up the lad, <u>and hold him in thine hand; for I will make him</u> a great nation.*"

Genesis 21:14-18

Chapter Twelve: Boys Will Be Boys

In all the areas commonly known to describe the behavior of men, there is one particular characteristic I want to look at. The title of this chapter may depict the bad behavior known to be displayed by many men, (so much so that it is often overlooked) of childishness. But that is not my focal concern right now. I want to look at the fact that men and boys for ages have been victimized through sexual crimes and, as most do, they never tell a soul.

Not being able to bear such a burden alone anymore than anyone else, they fall into the same prison of false identity as women often do. If prison walls were all one way mirrors, the sight of gang rapes, sodomy and torture could undoubtedly be witnessed regularly. Sadly, men are expected to be strong, tough and- - deathly silent.

Inside each of their souls a bad seed was planted. If it is not uprooted, it will grow up and around that man's very soul until he can't even recognize himself. I liken this to a wild ivy vine growing on a tree. It covers the tree so completely that it eventually prevents it from getting the sunlight and air it needs to survive, destroying it. Though there is no comparison between the tree's physical strength and that of the vine, the vine dominates by active progression, while the tree in its' slow and steady existence, stands strong, but helpless.

Emotions can run as wild as that vine. After being raped, the man often is not able to explain his feelings, control his thoughts, remedy his pain or shake his newly developed habits on his own. Eventually, he can become what God never intended him to be, whether it appears outwardly or not. If the tree is the man's heart, he cannot stand stoically as if nothing is happening. The choking vine has to be destroyed. Think about it, if the tree in this analogy represents the heart of the victim and the rape is the seed, then what is the vine?

Does God want his men to be hate-filled? Or fearful? Does God want men to seek vengeance? Is it God's intention for His men to feel so disconnected from His purpose that they begin to carry out the purpose of God's very enemy-even to the point of destruction? Or even worse still, becoming a rapist to someone else?

A bad seed is just that. Pain is pain; it knows no gender and it hurts no less. The power of evil is defeated only by the power of God...not time, will power, human strength or the appearance of it.

A man is still a man even if he has been raped. He is still a man if he cries and he is a man if he asks for help. God made him a man. No feelings of confusion, insecurity or inadequacy will ever change God's intention for each individual that He made. No circumstance is an acceptable excuse to decrease into less than what God created a man to be. Will a man be a man because he handles his trouble on his own? Or is he a man when he is wise enough to know when his ability is limited and he goes to God? Jephthah gave us a perfect example. He had been cruelly treated and humiliated, but it did not stop him from being all he could be or trusting God to handle his trouble.

A person caught breaking the law can bring up a whole lot of past pain as an explanation of the behavior, but that's too late. The time to bring up the pain is when the pain first comes. Otherwise, it will remain, dull, get ignored and continue to grow until it cannot be ignored any longer and destruction takes place.

There is an old trap that male victims and (some females) fall into. It is a cold, dark place that is too often overlooked. It is the mental place scarred children go to when they have been ravaged relentlessly, particularly the boys.

When a man sexually forces himself on another man in a show of sordid power, it deeply grieves the male victim's soul. It has been called "mental castration;" bringing a hidden sorrow that doesn't necessarily show up in sad expression. Anger, almost extreme at times to a point of rage is what seems to appear more. Discontent and continual drudgery. A malevolent attitude and bitter disposition. Sometimes, because he has been forced to sexually submit to a man, his pain even shows up as a blatant disrespect for women. Other times it appears in either a great fear of those in authority or, in sharp contrast, severe disdain for them.

It doesn't really need to be mentioned that abuse of substances like alcohol and drugs often tend to follow the man who has been sexually defiled. His shame is in essence far greater than that of a female because his body was used to perform what it was not even designed to do. The unnatural and emotional act brings a deeper level of confusion to the man as it leaves him feeling that he has reached a point of no return. His encounter leaves him in a 'No Man's Land'- still a male, but now with feelings that he is no longer a real man. He is furious that he has been forced into this altered state - - not a man - - not a woman, but in his mind, an unnatural thing...a monster.

He feels ruined, wasted and of no use. He cannot separate his aspirations from his memory so he tries to lose himself in a mental state where memory seldom appears. Drunkenness, wantonness, promiscuity, rage. All that strokes the ego enough to make him feel a little in control. He could not control the attacker, but he can control how he is seen thereafter...(or so he thinks).

The attacker observed him as weak. Now he will either deliberately show more strength, even brutally, or he will show extreme submission via homosexuality to snare other men attracted to him. Both moves give him back his power to a degree. He has to prove he is worth something or at least feel in control again. He feels he has to become either a man that no one will dare attack or even one that others will become weak by his seductive advances. Many times, however, he just won't care what happens to him and without a fight at all, becomes whatever he feels compelled to be.

This is all quite spiritual, but it is also physiological. The chemicals released in a man's body during moments of extreme emotion such as rage apparently override the constant ache his soul suffers from the rape. One cancels out the other and the lesser is forced to lie dormant, (at least temporarily).

Think of the men in the cities of Sodom and Gomorrah. Every single one of them both young and old came out to participate in a gang rape of the two visitors that came to see Lot. These men were so emotionally charged with rage that they were violent. This level of unified assault does not develop overnight. It is safe to assume that every member of that 'mob' experienced a similar induction into it, most likely in very early childhood.

To grow up that way, being subject to the maniacal demands of grown, angry and perverted men had to be a horrible way of life. So awful, perhaps, to see anyone else living free of it provoked a jealousy strong enough to do the same. *("...jealousy is cruel as the grave: the coals thereof are coals of fire, which hath a most vehement flame."* Song of Solomon 8:6) At the very least, living such a dreadful existence could have, on the other hand, instilled a fear of retribution for not participating in such an act of brutality.

Although society as we currently know it does not have the mob problem, there is still the secret suffering of male victims who deny being raped, standing, as that vine imposed- tree, as if nothing is going on.

But it is the man's denial that is the hold, that is, the unyielding, choking vine. Once the silence is broken, the vine can be destroyed. The rape was only the seed of the vine, planted by a cruel enemy. The vine is the secret denial of the rape. The way to becoming free of this is to realize that the secret is only held by you and the attacker.

In his drunkenness or mental obsession, the rapist knows what he did to each victim because rape, unlike a homicide, is never an accident. He may not even recall much about one victim or another, depending on how inebriated he was at the time.

In striking contrast, the victim can remember every detail, every smell, every sound, everything. Men, I've learned, would rather not recall it. Burying it in the back of their minds- suppressing it, so they won't suffer emasculation, (by memory) again and again.

Men are visual. Judging from the way men respond to other visual stimuli, I can imagine that if the memory of the rape is visually clear, and he is somehow reminded of it, he will react to the picture he sees in his mind much like he would in actuality. He might start to breathe heavy or not at all. His heart will race, he may sweat and even, just for a moment, mentally re-visit it.

Women endure these flashbacks as well, but with significant difference. Women react internally and the stress of it shows up in her physical health. A man at that moment, if he feels nothing else, feels weak and vulnerable and it's that feeling above all others, that depresses him into acting out. He doesn't get depressed into sadness and loathsomeness, necessarily. He, most likely, isn't going to sulk. If anything, he will go out of his way to avoid being 'prey' again. He never wants to appear weak or as an easy target to anyone. He especially doesn't want to be labeled effeminate. To avoid it, several outward expressions may occur in the male rape victim.

Hence comes the extreme body building, brutal sports, martial arts, oversized vehicles. Not to mention pursuing plenty of women, (the more naïve the better). A strong, outspoken woman is the last person he wants to be around.

There are some men who will in silence go another route. They stay down, becoming what they believe they have been reduced to. Their bodies were used in fashion as a woman's, so they in turn cower into effeminism, homosexuality and male prostitution. They turn, believing that they are not worth anything more.

Believing something is wrong with them, (especially if the rapist said degrading things to him) he may be deceived into believing the attacker's lies, particularly if he was of a mild nature before. Others still never act on their feelings and lead 'normal' lives outwardly, while struggling indefinitely on the inside.

Other men seek endlessly to affirm their manhood by becoming obsessed with sex and the sexuality of women. Pornography addictions get out of control to the point where no sex act is too grim- and no satisfaction is ever reached. He becomes a whoremonger.

The feeling of being ruined drives the next group over the edge completely. Not able to live with the death of their ideals- they veer towards total death with suicidal tendencies.

The worst lot of all is the victims that in turn, victimize others. Either to prove their manhood or punish the weak for allowing themselves to be weak, these men have definite agendas. The torture of it all is the restlessness before and the horrible darkness afterward.

After more than two decades of study, interviews and observation, this is only a brief generalization of the psyche of a male victim of rape. However, I am sure it is inconclusive. The same enemy of us all will try anything – anything to keep the victim from where he is destined by the Lord God to go. (See Jeremiah 29:11-12) God's plan for every individual is one of goodness and blessing. The enemy of God wants to thwart that divine plan by tossing in an explosive situation or occurrence to, if it were possible, block the path to their destiny. But only the person walking the path chooses to continue on it.

Being silent will never be the way to get over the pain of what happened. They say that boys will be boys. They're not supposed to talk about "this stuff". They don't share because the risk of being labeled for life is too much to imagine. No, historically, boys handle it their own way, unfortunately. But it doesn't have to be like that.

We have learned so much about David already with regard to his behavior as an adult, but what from his childhood can we learn? In reading about him, we know that he was the youngest of several brothers. He was a shepherd boy who was also a talented musician and lyricist.

For some reason, when the prophet of God went to David's father Jesse's house to anoint the next king, his father didn't even consider David. In fact, he did not present him until after he had brought the others first and then only after the prophet insisted that he bring all of his sons to him. We don't really know if he didn't think much of David or if he just didn't want David involved. But it's interesting, isn't it, that he didn't want to see his own son as king? And what father would allow his youngest child to work alone where dangerous animals lurked? Then to send him unarmed into a battlefield to take food to his older brothers, who were warriors, instead of a servant…

really makes one wonder if David was valued in his father's eyes. Had he hoped that David would be harmed and therefore make a way for one of his favored sons to be crowned?

Then there were David's brothers, who actually scolded him for coming to the battlefield and instead of thanking him for risking his life for their well-being, called him names. Note the disrespect – this was after David had been anointed to be the next king right in the presence of his brothers. Later we look at the fact that David had to literally live as a fugitive from the current king who, out of envy wanted to kill him.

So, there was the child David, very good looking, a loving shepherd, talented in music, strong in faith towards God and brave – despite being on the run for years. Later, his own wife disrespected him for how he worshiped God. If that wasn't bad enough, she later was taken from him and given to another.

Would a man like David, with so many emasculating situations in his life feel the pressure to prove his manhood? To be so disrespected and dishonored living as if he was a coward to escape from a mad king, not to mention having a wife that belittled what mattered most to him, what kind of pressure would that cause a man?

All of this may have been silently on his mind for years, eating at David until the day he decided to take full advantage of his privileges as a king and literally take a woman for his own personal pleasure, despite how good God had been to him. Boys will be boys.

"...because [you have not only despised My command, but] you have despised Me and have taken the wife of Uriah the Hittite to be your wife..." *2 Samuel 12:10 (Amplified Version)*

Suffering in silence can be a very dangerous state of mind. David's heart was open to God, but it was that mind that needed to be restored and according to David's writings, God did. (Ps. 23:3a) The psalms that David recorded are evidence that he did open up to God and share all of his feelings. It may surprise you to find that many of his thoughts are a lot like your own. David turned completely around and became the greatest earthly king known to man, but only after he stopped despising God and opened up his heart to Him.

Maturity comes when the male decides to grow up past the boy-mentality. I know that men innately, are problem solvers. Men deal with situations in 'answer mode'. They don't wish problems away, talk them away, explain them away or deny them. They look for the answer and do what needs to be done to fix the problem. Otherwise, they seek out someone who can.

The only problem with that mentality is that rape cannot be remedied by any human effort. The only help is in Jesus Christ. He is the way, the truth and the life. No man will ever come to know the One who is the answer unless they go through Christ first.

That's going to mean breaking the 'men don't talk' code. Honestly, enough suffering has already taken place. Open your mouth. Ask the Lord to heal and restore you. Ask our Heavenly Father to father you. Ask Him. He will do it. He will. You may not be able to talk to anyone else right now, but talk to the Master. He can make it all better, that is His promise.

"Come unto me, all ye that labour and are heavy laden, and I will give you rest. Take my yoke upon you, and learn of me; for I am meek and lowly in heart; and ye shall find rest unto your souls. For my yoke is easy, and my burden is light." Matthew 11: 28-30

Give your hurts to the Lord. How? Praying a simple sentence, asking for His help. Call on the name of the Lord and you will be saved. He can save you from this (seemingly) doomed existence and deliver you into a life, on this earth, mind you, of peace and calm. He can change everything.

"He leadeth me beside still waters." Psalm 23: 2b

The attack may have left you not really knowing what to do next or where to turn. But God wants to show you how to return to your path of destiny. How to walk upright again--focused on your purpose. He has availed himself to you. You have to make the next step. If that means allowing yourself to talk to God through tears, then do it. If getting to Jesus means pouring out your soul in a miserable heated course of questions, then ask away. Your emotion won't make God turn from you. Whatever height of anger or depth of despair you find yourself in because of the attack, God will still be willing to receive you.

"For I am persuaded, that neither death, nor life, nor angels, nor principalities, nor powers, nor things present, nor things to come, nor height, nor depth, nor any other creature, shall be able to

separate us from the love of God; which is in Christ Jesus…"

Romans 8:38-39

I do strongly advise you to go to a Minister or a Christian counselor or even someone you know that is a Christian. They can help you find peace in God. " *…in the multitude of counsellors, there is safety." (Proverbs 24:6)* Whatever way you decide to get to Jesus, just get to Him anyway you can. He can work everything out.

Let's remember Jephthah. He was the son of a man with a dark past. Because his mother was a harlot, there was great strife in his household. But regardless, and in spite of how his own family treated him, he determined within himself that he was going to press on and make something of himself. He didn't wait for someone else to affirm him or give him worth. He knew who he was.

Males many times either don't want to talk about their issues or on the contrary, some want to only talk. The wise way to handle any issue is to talk only to those who will offer true godly wisdom and then adhere to that counsel.

True wisdom comes when we follow the ways of the Lord. Truly this is the only way to find the path to our destiny. Jephthah was a man who relied on the Lord. His family let him down even though, as we read later in the account, they knew who he was. They chose not to see his worth until they desperately needed him. Yet in spite of them, Jephthah was a man of prayer. Take his example; no more silence and denial. Pray for your life to be restored. God will do it, He will! I am certain of that.

Journal Entry About God's Power to Recover Sight to the Blind:

Family & Friends

Sometimes, we simply forget who we are. As family members we hold a high position in the lives of our loved ones. A rape has the capacity to make us lose faith in who we are supposed to be. We are meant to be great, to do great things. To change the world and leave it better than when we entered it.

When you see that your dear daughter or son is living beneath their potential since being assaulted, you must help them to find their way back. The enemy mistook them as some weak soul to be used and abused - -but God says not so.

You have to help them believe in their value. Encourage them to discover or re-visit their hidden areas of talent and skill. Help them find their passions again. Memorize bible verses that speak of God's demonstrative love. Don't give up on her or him no matter what.

If she has become pregnant as a result of the rape, it is vital that you consider the preciousness of the life that is coming into the world. Of course this was not the ideal plan, but the past is passed—help her to get a new plan. The assault was bad, but there is nothing bad about a child.

Try to make her comfortable without ignoring her emotional needs. She has two-life changing events to deal with. Be careful to listen to her so that you know which area she may momentarily feel troubled about.

Don't be afraid to start a conversation with her. It is doubtful that she isn't constantly thinking about one thing or the other. Just be ready to listen and encourage.

Don't allow her to slip into a state of perpetual sorrow. Be determined to keep her alive. Help her to stay healthy physically and emotionally.
You can do it, there are lives depending on you to do so.

section five:

"...to preach <u>deliverance</u> to the captives..."

"And they came over unto the other side of the sea, into the country of the Gadarenes. And when he was come out of the ship, immediately there met him out of the tombs a man with an unclean spirit, who had his dwelling among the tombs; and no man could bind him, no, not with chains: Because that he had been often bound with fetters and chains, and the chains had been plucked asunder by him, and the fetters broke in pieces: neither could any man tame him. And always, night and day, he was in the mountains, and in the tombs, crying, and cutting himself with stones. But when he saw Jesus afar off, he ran and worshipped him, and cried with a loud voice, and said, What have I to do with thee, Jesus, thou Son of the most high God? I adjure thee by God, that thou torment me not. For He said unto him, Come out of the man, thou unclean spirit. And He asked him, What is thy name? And he answered, saying, My name is Legion: for we are many. And he besought him much that he would not send them away out of the country. Now there was there nigh unto the mountains a great herd of swine feeding. And all the devils besought him, saying, Send us into the swine, that we may enter into them. And forthwith Jesus gave them leave. And the unclean spirits went out, and entered into the swine: and the herd ran violently down a steep place into the sea, (they were about two thousand;) and were choked in the sea.

And they come to Jesus, and see him that was possessed with the devil, and had the legion, sitting, and clothed, and in his right mind: and they were afraid. And they that saw it told them how it befell to him that was possessed with the devil, and also concerning the swine. And they began to pray him to depart out of their coasts. And when he was come into the ship, he that had been possessed with the devil prayed him that he might be with him. Howbeit Jesus suffered him not, but saith unto him, Go home to thy friends, and tell them how great things the Lord hath done for thee, and hath had compassion on thee. And he departed, and began to publish in Decapolis how great things Jesus had done for him: and all men did marvel." *Mark 5: 1-20*

Chapter Thirteen: The Man in the Tombs

The man who lived in the tombs was a man that was possessed by many devils. He was all but consumed. The devils within him had driven him to a point of such madness that his community tried to tie him down and bind him up. Their efforts were futile however, as the spirits caused him to be violent and uncontrolled. He cut himself with stones continually and cried loudly all night. In addition to this torture, he was forced to live in a cemetery.

Aside from the thought of how awful it must have been to witness someone living like that, just imagine what it was like to hear a grown man crying day and night. At some point it had to tear at the hearts of the people to be around him. Obviously, the people could not help him. Their attempts to detain him were of no use. No one really knows if their efforts were seated in love and concern for him, or just base fear of the poor man. His life was one that was gone to waste, or, so it seemed. They wanted to control him or perhaps keep him from hurting himself or possibly others, but his internal bondage was far greater than anything they could contain. That is how people with such problems back then were handled. Actually, society today still tries to manage such suffering people with physical bondage when internal freedom is what is needed. Only God can deliver true liberty to the soul of a person.

That's why no one is helpless. It didn't matter how far-gone that man was, God was still God and could therefore save him. Here was a man that was isolated in a dead place, crying night and day, hurting himself and suffering constantly. The thing that stills my heart that we cannot ignore is the fact that the man was crying. He was still a human being. In spite of the wickedness that flooded his soul, he still had a heart. It was that heart that made him cry out to Jesus and from that heart he worshipped Him.

When Jesus came near to where the man was, he ran to Jesus and fell down and worshipped him. His body may have been possessed, but his spirit was still alive enough to cry out to the Lord and worship him. But right after that moment, the devils began to speak to Jesus of their own interests. Immediately recognizing Jesus, and knowing their place and fate, they boldly made a request for mercy. (It's sad that we find ourselves too afraid to ask the Lord for what we desire.)

We can't compare their requests to our prayers; theirs was based on legality, ours is based on hope. Their requests, interestingly enough, weren't well thought out, otherwise, they would have realized that not even pigs wanted to be in-dwelt by them. So they still ended up cast out.

How did the man become possessed in the first place? Devils can't just take over a man's body randomly as they choose to. They have to be allowed access. Witchcraft, sorcery, drugs, the occult, psychics and mediums are but some avenues where the devil can gain access. That is why the Lord forbids us to engage in such activities. Many people that seek answers from dark sources such as these are either curious, seek vengeance, feel powerless or want more control of their lives. Believing that they will be better off drives them to step out into dark places and side with the enemy's enticing plea. No one ventures in this unholy path to be made worse off, but in doing so, they veer away from the Lord God, and end up worse off anyway.

What if this man sought out the promises of dark views to answer questions he had? Perhaps he felt powerless for some reason and wanted strength. How do we know that he wasn't sexually abused as a child and turned to such a dark venue to escape his suffering or even seek revenge? Nowadays many children's stories glorify dark powers to make children feel less helpless. The only problem is that these impressionable minds are indirectly exposed to witchcraft, sorcery, wizardry, necromancy and the occult, making them accepting of what God, for our own sake, forbids. (See Deuteronomy 18: 9-14)

How can anyone know what drove that man to a point where he was taken over so horribly? We don't and never will know. All we need to understand and keep in mind is that this man was about as far-gone as any human can possibly be on this earth...yet Jesus helped him. Do you have any idea why his story is so significant for us today? Just think about it, you are not even close to being as bad off as he was, and if Jesus could deliver him, set him free and restore him to his right mind again, you can be absolutely sure that He can and will do the same for you.

Understand, this man was not a monster, he was just a man with an overwhelming problem. Try to see the humanness in him. It is possible that the man stayed in the cemetery to keep the spirits inside of him from driving him to hurt someone. Perhaps he didn't want innocent children to be frightened at the sight of him in that condition. When Jesus told him to return to his friends, it is clear that this man had a 'life' before the spirits possessed him. He was a man that was loved and knew how to love, just like you and I.

We don't know how his bondage began, God left that part a secret from us. He only allows us to know what we need to know, and that is the fact that God is the answer no matter how our souls became bound. Or to what level of severity we may be emotionally enslaved.

So don't look at yourself in the condition you were left in after the attack. You are not a poor, unfortunate soul. You are not hopeless. You are not a victim! You are a person of power and potential by virtue of the very fact that you were created in the image of God. How does the Lord look at you? He sees you as precious and valuable. He sees you as strong and capable. He sees you as His very own.

How do you see others? You have to look at them the way the Lord does. Don't give up on someone just because they were violated. Don't dismiss them as pitiful, your pity is not helpful. Pray for their recovery and treat them as the living breathing creation of God that they are. We don't have time to let our fellow humans stay in a place of agony, crying day and night, unable to receive real help from anyone, bound by man's definition of them. Pray.

At times, a victim is alive, but left with something not possessed before. In addition to fear, shame, anger, hatred, bitterness or confusion, many victims find themselves struggling with a trace of the assailant's internal struggle. In particular, the feeling of powerlessness, insecurity or disrespect for sex and gender differences.

It may be immediate, or even years may pass, but in many instances victims find themselves doing or thinking acts that are uncharacteristic. This may surface as homosexuality, promiscuity, voyeurism, rage, gender-bigotry (chauvinism), prostitution, substance abuse, or exhibitionism.

But no matter how bad it is, God is the answer. God didn't cause the pain- the enemy of your soul did. God is the answer to your problem. He is. What do you need? Comfort? He is our comforter. You need peace? He is our peace. You need your mind stabilized again after all that trauma? He is the restorer of our soul.

Cry out to the Lord, that is what that dear man from Gadara did. He cried and cried for God only knows how long. I believe with all my heart that Jesus, regardless of the stigma of a holy man doing so, made a special trip to that cemetery just to respond to his cries. God is that good. I also believe He is coming to answer yours. So wherever you are right now, cry out to Him. It's okay to cry for help.

Tears of grief and sorrow have a place and serve a purpose for a time. But when you finally see that there are some things that must change, namely the fact that the sorrow is still there and still strong and still very emotionally draining, you need to call to the Lord to help. He will wipe your tears away and give you a joy that you won't be able to explain away. He alone can restore your soul, so cry out to Him. There is nothing the Lord won't do to answer you when you call Him with your whole heart. God can deliver, but we must seek Him to do so. Nothing is impossible for God and there is nothing too hard for Him to do. Remember that.

Before we close this chapter we have to mention the possibility of a different kind of sorrow. In the last section we looked at the worth of children, even children that were conceived as a result of sexual assault. We mentioned the enemy's attempt to make you want to abort the child to rid yourself of any residue of the attacker. For some of you, that message needs to be amended. The reality is, some of you may have already succumbed to the dread of the whole experience and terminated the pregnancy.

The message for you is this, God can <u>still</u> restore your soul. There is no one who is too far-gone in God's eyes. You can always come to Him. God is loving, merciful and gracious. He is willing to forgive. The only sin God will not forgive is the sin you don't repent of. So your cry is different; yours is a plea for forgiveness and healing. Yours is one for help to recover from that shame that you participated in. God is graceful and forgiving even if you succumbed to the lie of abortion.

When the enemy attacked me with attempted rape the third and final time, I was seventeen years old. As I mentioned before, the man stopped abruptly and quietly and let me go. What was strange was the aftermath as my body underwent so much shock that it changed. I couldn't eat, I was ill every morning, stayed unusually bloated and had no menses for nearly three months. Not being sure of anything, the adult females in my life tried to explain to me and even prepare me for the possibility that perhaps I *had* been penetrated, if only slightly, and thus conceived.

I cannot tell you the grief, trauma and insurmountable disgust I felt at hearing this. I was so confused I didn't know what happened anymore. My mind raced with questions. Was I pregnant? I was very sore at first, but I thought that was from the sadism, not - - himself. Were these ladies right? What did I know about all of this anyway? Had I fainted while I was with him? I had passed out a few times the year before from heat and so forth, so it wasn't completely unlikely. Was I crazy, I mean, I thought I was sure about everything? But then, I did refuse a medical exam for fear of more shame after the attacks, so, I really wasn't completely straight on what was going on.

All I was sure of is that my body was not normal and if there was a child, I did not want anything to do with that man's baby. He disgusted me so...but God said "Thou shalt not kill...", so what was I supposed to do? For the first time since I even began to understand what these matters were about, I had a difference of opinion.

So there I was, day after day, week after week, hoping for what I used to dread and dreading what I used to dream of experiencing. What a messed up state of mind I was in. If that were not heavy enough, these precious ladies, still trying to be prepared, often asked me, ever so gently, if my cycle had returned. The stress of everything was hard for all involved. In all of this, my thoughts were not, I honestly admit, very much on the things of God, other than to just cry "Why me?" I was simply scared out of my young mind.

At first, I wasn't going to share this with you, but, I want you to be free and if I have to bare my very soul to you to help usher in your liberty and healing, than so be it. God bared all for me. Although I was not pregnant and by examination later learned that my hymen was intact, I was still completely shocked at how my heart towards abortion had changed…or had it? No. My heart was still my own, it was my mind that was in turmoil. I was completely divided within my self.

"For the word of God is quick, and powerful, and sharper than any two-edged sword, piercing even to the dividing asunder of soul and spirit, and of the joints and marrow, and is a discerner of the thoughts and intents of the heart." *Hebrews 4:12*

Believe it or not, I, the one that was called to the ministry just a few months prior to all of this and had debated fervently against abortion, would have done anything to be washed of that horrible feeling. I just wanted to forget it ever happened. To have to explain to everyone that I was pregnant and by whom, was more than my imagination could handle. Even though I adored the Lord, I didn't know what to do. My heart and mind were completely divided. I knew what the Lord said about killing, but my state of mind was so altered, I couldn't put a sensible thought together. Walking around pregnant by him seemed unbearable.

When my body finally did calm down from the stress of the attacks, I don't recall feeling relieved, by then I was pretty numb. I was so ashamed at my pattern of thinking, that is, when I did allow myself to think about it, I was weighed with the guilt of it all. I wish I knew then what I know now. Even for years afterwards, I wondered, if there was a pregnancy to deal with, what would I have done? This, my friend, is how we are alike. If you did abort the child that was conceived by rape, you and I, because of my resolve to do so, are very much the same. Can you believe that I carried guilt for years because of that? But, thank the Lord, the same Word that He spoke to me when I prayed applies to us all:

"There is therefore now no condemnation to them which are in Christ Jesus, who walk not after the flesh, but after the Spirit..."
Romans 8:1

God has provided whosoever will with a chance for a new start and a clean slate. Whoever you were before and whatever you did before are all removed from eternity's record once you give your life to God and surrender to His will and His ways. If you just come clean before the Lord through faith in Jesus Christ, confessing your sin and turning away from sinful living, He will give you a new life.

"Therefore, if any man be in Christ, he is a new creature; old things are passed away; behold, all things are become new. And all things are of God who hath <u>reconciled us to himself</u> by Jesus Christ, and hath given to us the ministry of reconciliation." *2 Corinthians 5:17-18*

"But what saith it? The word is nigh thee, even <u>in thy mouth, and in thy heart</u>: that is, the word of faith, which we preach; That if thou shalt confess with thy mouth the Lord Jesus, and shalt believe in thine heart that God hath raised him from the dead, thou shalt be saved... For the scripture saith, Whosoever believeth on him shall not be ashamed..."
Romans 10:8-9&11

If you will believe it, your life can begin again right now, no matter what happened. Tell the Lord that you want to be changed. You know what lies in your heart. I can't put the words together for you, I can only encourage you to open your mouth and cry out to God. He heard that miserable man that lived in tombs, he heard me and He will hear you. Don't let your heart stay in a dead place, grieving over the death of the child or perhaps even the shame for your lack of grief. Just let the Lord come in, make the devil release his chokehold on your soul and return you to who you are supposed to be. Live after God's way, free from guilt and shame. Walk in the light and remember, the child is in the arms of God now, out of the way of the world's evils. Be at peace.

The Blessed Children

All the pains this world can name
The trials and sorrows
That all here go through;

The pain of rejection, fear, shame
These blessed children
Never knew.

No ongoing torment
Brought by fear,
No loneliness, no hurt or grief;

Not one will shed a tear
feel a shiver of cold-
or faint from heat.

Never a need to scream
Will arise
for terror they won't ever know.

No melted heart or broken dream.
A saddened face-
They'll never show.

Only a peaceful, joy-filled fate
Of abiding forever
With God above…

And a sweet anticipating wait
To reunite with
all earthly loves.

Chapter Fourteen: What Was That Sound?

Oppression is a child of fear. It depends on it and feeds on it. Without fear no one can be oppressed. The day you refuse to be afraid, is the day you are free from oppression. It isn't right to be oppressed or in fear. Making things right with God also means that you submit to His protection by giving Him your trust. Once you trust Him, truly trust the Lord, you are resisting the fear and it will leave you, taking oppression with it. *'...thou shalt be far from oppression; for thou shalt not fear...'* (Isaiah 54:14) Before I came to my place of wholeness, I had to recognize the hold that my fears had on me and learn to trust God to deliver me from all of them.

"I sought the Lord, and he heard me, and delivered me from all my fears." *Psalm 34:4*

Before fear overcomes you, learn and understand the truth about fear. It is a spirit that God never intended you to embrace. It will torment you mercilessly and continuously. It will wear on your soul until you are too weak to stand up when you really need to. You must come to understand that God provides specific resources to make you able to stand. Begin to read what the written Word of God says about fear and read it aloud to strengthen its power in you. Think earnestly on it as often as you can, especially when you feel anxious. Let the Spirit of the Lord give you understanding and insight.

If you were on a ship, what would be the safest place to be when a storm is about to come? In the heart of the ship, deep below. Likewise, you must learn to go deeper in God by going deeper into His word in study and observant contemplation. Be more intense in your prayer and involvement in church and the Lord's service. Strengthen yourself by embracing God's power, love and soundness of mind. You do not have to live with fear, it is a terrible occupation to hold.

When lions attack wild cattle, there is a system to how they accomplish the feat of bringing down such a massive animal that can easily gore a lion to death. Working in a pack, the first lion must secure a hold on the animal's back. The second lion holds the head in place by clenching the beast's sensitive nose to control its dangerous horns. If the lion holds on long enough to this area, the beast will yield. Finally the lions continue to work until they can twist the head to seal the windpipe of their prey.

As phenomenal as the cooperative effort of the lions is, what is more amazing is that the beast is brought down by the lion pack right in the presence of the cattle's herd. The fear in the wild herd is so great, they would rather look out for their own protection instead of joining forces to protect the one being attacked.

If fear can paralyze a beast that weighs hundreds of pounds more than it's predator, then fear must be something to reckon with. We, on the other hand, are not beasts. Therefore, we cannot allow ourselves to ignorantly allow fear to stop us from defeating the enemy of our soul. Because of Christ, we are already conquerors.

"...Nay, in all these things we are more than conquerors through him that loved us..." *Romans 8:37*

We simply must not let fear imprison us. With everything that happened, it is quite understandable to have to deal with fear regularly. Some of the scariest moments for some is when the house is entirely too quiet and everyone is asleep, except maybe-him. What was that sound? Did you hear something? Shhh...be still...

It's not the fact that a sound was possibly heard, it is the fear of what or who may have caused the sound. The sound may be as strange as an unidentified thump or a doorknob being slowly turned when the rest of the household is asleep. Or the sound of your own heartbeat pounding as you try to sense if someone is standing in the darkness. It may even be the eerie feeling when there is no sound at all. (Silence isn't always necessarily golden.) Nonetheless, fear seems to come in right after that sound gets your attention and that's where the suffering begins. The sudden cautious and steadily gradual feeling of danger is not merely an emotion, it is a reaction to the presence of fear, which, is a spirit.

'For God hath not given us the spirit of fear; but of power, and of love, and of a sound mind." *2 Timothy 1:7*

That is how the Word of God describes fear, as a spirit. If you think you are overacting, you aren't. That spirit comes just as any other evil spirit does, to kill, steal and destroy. So, to a certain extent, your reaction to want to preserve yourself is merited. The spirit of fear cannot literally murder you, but it can torment you to the point where you don't want to live. It can steal your peace and destroy the 'moment of contentment' you enjoyed just before. Or worse, drive you to a place of self-destructive behavior.

It does this by instantly and continuously reminding you of your past pain and devastation. It will use something as intangible as a sound to make you wander in the wilderness of yesterday's distress. Unstopped, the spirit of fear can beguile you into being it's prisoner, trapped in your own home, paralyzed from moving on in your life.

But God's way is not for you to be a prisoner but an overcomer. Moving forward into your destiny, accomplishing what God created you to achieve. The greater your purpose, the greater the threat to the kingdom of darkness you are.

When I was in my twenties I lived alone for a few years. I can vividly recall sudden moments of terror that came over me. Each episode was initialized by the most harmless occurrences. A windy night. Strange voices in the apartment building hallway. Footsteps outside at night. (It didn't help that I have always been blessed with a keen sense of hearing.) Being alone when the maintenance man came to fix something. A glare, a stare or even a compliment by someone who meant well. Being the only female on a subway car with several men on board. (Each minding their own business by the way.)

The situation didn't really matter. Years passed by and the feelings would still come. The nightmares still happened. My heart still raced at the very thought of being helpless at the mercy of someone else's hand.

As stated before, that feeling of danger is not necessarily an emotion, but a reaction to the presence of fear. The more that spirit is allowed to direct you, the more it takes over your life. You must be rid of it. The Bible says that God didn't give us the spirit of fear so if it isn't from God, then it is not *of* God. Therefore, if it is not of God, it must be resisted. How? By doing the opposite of what that spirit is attempting to 'drive' you to do. I mean this literally.

Think about it this way; you wake up after a horrible dream. (I've done this many times, so I know). Once you are aware that it was a dream, you notice that the room is darker than usual. Your heart is racing. You are almost sure that someone is in the room. You need (or at least you really want) a light on. After you have calmed down, maybe after a glass of water, you replay in your mind every scene of the nightmare and hope you don't see it again. It could be a very long night (well lighted, no doubt), but long.

How many restless nights will you have before you stop that enemy from stealing your peace? For years I did everything I just described. Society made these actions seem normal, but yet society fails to tell us that giving in to fear continuously like that will destroy us.

God revealed to me through His written word, that I don't have to suffer from the torment that fear brings. His very spirit lives inside of me, comforting me as well as leading and guiding me into all truth.

"There is no fear in love; but perfect love casteth out fear: because fear hath torment. He that feareth is not made perfect in love..."
I John 4:18

God was faithful enough to let me know when I needed to be cautious and watchful or when I could just relax. I remember one night when I lived alone in Oklahoma, I was on my way home from work and I felt like I really needed to be home. It was as if the Lord was urging me to get home immediately. Once I got in, that 'feeling' passed. It wasn't fear, it was a spiritual awareness. See, because God gave us His Holy Spirit to be with us, we have inside information about what is going on in the invisible spirit realm. Not foretelling, but the wisdom and knowledge given by God Himself. Remember earlier when I talked about being on the offensive as the best defense? Its preparation, not reaction, that sustains us many times.

That night, as I settled in, I drew a bath, ate dinner and got ready to enjoy one of my favorite past-times: old movies. Right before the show started, I decided to run to my neighbor's apartment to borrow something for the movie. I don't remember what it was I wanted, I think it was oil to pop my popcorn or something. I do, however, remember the shock I felt as I ran right into the middle of a gathering of young men. There were at least a dozen on both sides of the parking lot and the two groups faced each other with a strange silence. Some of them had shining chains, others had bats…it was all overwhelming. My quiet little, all adult garden apartment complex now looked like a movie scene itself. But I was too far outside to turn around without possibly raising suspicion. So I slowed my run to a brisk walk and went to my neighbor's anyway.

I waited there prayerfully until the group seemed to disperse. By now I was quite open to God's leading. I realized that He was directing me from the beginning. Just because I didn't feel fearful before didn't mean I should have ignored what I knew inside--which was the fact that I needed to be home. I never did find out what the commotion was about out there, but it didn't matter, God taught me a lesson. When He is leading us, he doesn't torment, worry, scare or over-excite us. He doesn't give us an anxiety attack to get our attention. The spirit of fear, on the other side, will leave you breathless, spit-less, covered in sweat and still without clarity of direction or what positive action to take.

We must learn the difference. The spirit of God through His loving and sometimes mysterious ways will alert you to danger. The spirit of fear will have you running from something that only exists in your imagination.

God's holy Word will make the lines of differentials clear. Study what it says about fear, courage, vengeance, protection and valiance. Learn about God's protective measures towards us. Memorize His words, for they will light your way the next time you wake up in a dark room from a bad dream or a strange sound.

Recite them to yourself before the enemy tries to attack you and you will be fully prepared to speak those very words if that time comes again. *"Be sober, be vigilant, for your adversary, the devil, walketh about as a roaring lion, seeking whom he may devour."* *(*I Peter 5:7*)* Notice I highlighted the word *may* here. A lion will not hunt one that is strong, but one that is weak. So—don't allow yourself to become weak.

I do believe that the man from Gadara that was so plagued with devils had some level of weakness in him that allowed him to be taken over so brutally. There is no way a Christian man that is in

God and filled with His spirit can be possessed by a devil, though his may be persuaded by one. Therefore, weakness is not a tolerable option. Take the Word of God as you would food. Keep putting it in your spirit until you are strong. The day will come when you will be so empowered that the enemy won't even attempt to hurt you like that again.

See, now if I have a nightmare, I don't cut the lights on. I don't stare into the darkness and try to wonder what the shadows are. I don't get up and pace the floor and I don't call a friend to console me. God's words have made me strong. There is enough of His Word in me for the Holy Spirit of God to just bring to my remembrance what scripture I need to use as my weapon at that time. *"But the Comforter, which is the Holy Ghost, whom the Father will send in my name, he shall teach you all things, and bring all things to your remembrance, whatsoever I have said to you."* (John 14:26) *I speak forth those very words and pray and worship God for always being with me. I don't give fear a chance.*

Now the only sounds I wonder about are the ones worth my time... that is the ones the Holy Ghost directs me to take time to investigate. I don't even raise my eyebrow at the rest.

Knowing the truth of God really does make you free. I don't have to try to prove I'm not afraid to live without fear. I have just learned to listen to God's mighty Spirit first. If you can do that, you will be so much better off. You too will be- - free. Never forget the man from Gadara. When Jesus freed him, he was so free that his very appearance changed. No longer pacing the cemetery, but sitting down, clothed and in his right mind. Soon you too, will be completely restored. No longer crying night and day, but free.

"If ye continue in my word, then are ye my disciples indeed; and ye shall know the truth, and the truth will make you free."

John 8: 31a-32

Irreclaimable

The night is finally over
The stark shadows are fading even now

Silence, so strong you can feel it,
broke with the continuos
chirp of the lark

"What, tell me, what is there
to worry about?" And how is it I feel no press?

The dawn breaks as I
ponder this new freedom.

Wait, is that a lark? I don't know what bird that is-
I just can't believe that I actually hear it.

Who knew horror was deafening?
Blinding too, I guess. Thank God, He can fix anything.

Anyway, I am no more bound to a perpetual night.
No more am I captivated in the dark.

I do not belong to it, no, not anymore-
"I am forever, irreclaimable."

Chapter Fifteen: He Is Risen, Are You?

When we read the account in scripture of the life of Jesus Christ, we cannot help but be moved and astonished at the remarkable manner in which He died. No other event in history was as extraordinary. Even history's records are divided around His life and death in the terms B.C. (Before Christ was born) and A. D. (After His Death).

There are several significant occurrences that took place before He died that need to be elaborated on here. The reason is that I am hoping you will see in the comparison great similarity to what you may have suffered yourself. Jesus went through an awful lot for us. It helps to know that He alone truly knows how we feel. Let's look at the account recorded in the gospel of Luke of the moment when Jesus was preparing to give His life for us and then study the chart to see the relativity of the suffering of a victim today.

"22:44*...And being in an agony he prayed more earnestly: and his sweat was as it were great drops of blood falling down to the ground.*
45*And when he rose up from prayer, and was come to his disciples, he found them sleeping for sorrow,* 46*And said unto them, Why sleep ye: rise and pray, lest ye enter into temptation.* 47*And while he yet spake, behold a multitude, and he that was called Judas, one of the twelve, went before them, and drew near unto Jesus to kiss him.*
48*But Jesus said unto him, Judas, betrayest thou the Son of man with*

*a kiss? [49]When they which were about him saw what would follow,
they said unto him, Lord, shall we smite with the sword? [50]And one
of them smote the servant of the high priest, and cut off his right ear.
[51]And Jesus answered and said, Suffer ye thus far. And he touched
his ear, and healed him. [52]Then Jesus said unto the chief priests, and
captains of the temple, and the elders, which were come to him, Be ye
come out, as against a thief, with swords and staves? [53]When I was
daily with you in the temple, ye stretched forth no hands against me:
but this is your hour, and the power of darkness. [54]Then took they
him, and led him, and brought him into the high priest's house. And
Peter followed afar off...*

*[63]And the men that held Jesus mocked him, and smote him. [64]And
when they had blindfolded him, they struck him on the face, and
asked him, saying, Prophesy, who is it that smote thee? [65]And many
other things blasphemously spake they against him. ...*

*[23:1]And the whole multitude of them arose, and led him unto
Pilate. And they began to accuse him...*

*[10]and the chief priests and scribes stood and vehemently accused
him. [11]And Herod with his men of war set him at nought, and mocked
him, and arrayed him in a gorgeous robe, and sent him again to
Pilate....*

*[20]Pilate therefore, willing to release Jesus, spake again to them.
[21]But they cried, saying, Crucify him, crucify him. [22]And he said unto
them the third time, Why, what evil hath he done? I have found no
cause of death in him: I will therefore chastise him, and let him go.
[23]And they were instant with loud voices, requiring that he might be*

crucified. And the voices of them and of the chief priests prevailed.
[24]And Pilate gave sentence that it should be as they required. [25]And
he released unto them him that for sedition and murder was cast
into prison, whom they had desired; but he delivered Jesus to their
will. [26]And as they led him away, they laid hold upon one Simon,
a Cyrenian, coming out of the country, and on him they laid the
cross, that he might bear it after Jesus. [27]And there followed him a
great company of people, and of women, which also bewailed and
lamented him...."

Jesus Pre-Crucifixion Experience	Common Experiences of Victims
1. Disappointment (verse 45) for being left alone without real support.	Many times the victim is not supported by the ones who should be there, though unintentional, it is always disappointing when the victim is left to deal with this alone.
2. Jesus was in Agony (verse 43)	This state of being is exactly what a victim experiences.
3. Someone he cared for betrayed Jesus. (verse 48)	Many times a victim is assaulted by someone he/she knows and even trusts.
4. Violent weapons were used to threaten Jesus (verse 52).	Rapists use weapons to put fear into the victim and to ensure surrender to their will.
5. Jesus was seized and led away (verse 54).	Being seized is not just being physically taken, but also when the soul is seized through devastation and led away to a place of internal grief and torture.
6. Jesus was treated with contempt, scoffed at, ridiculed and beaten (verses 63 & 65)	This tactic of demeaning words and ridicule is aimed at bruising the soul and making the experience more devastating and lasting than the physical attack itself. Rapists use degrading words to hurt the victim on a deeper level.

7. Jesus was accused and tried unjustly (23:10)	Sadly, the reason many victims do not report the crimes committed against them is to avoid the mortifying, slanderous, verbal assault that will most likely come against them.
8. Jesus was forced to wear mocking clothes to be further humiliated (23:11)	Anything the attacker can do to humiliate, demean or degrade the victim before others make him feel more superior.
9. Jesus was sentenced with the same punishment as what a hardened criminal would have received. (23: 23-25)	Solitary confinement, the most dreaded punishment besides execution, is how the enemy of our souls tortures the victim; using the victim's own mind as a prison.
10. Those who loved Jesus were grieved for Him (verse 27).	Being ravaged relentlessly affects everyone that cares for the victim. The younger the victim or the more brutal the attack, the more the loved ones are crushed.

Do you see the parallel? Jesus endured all of the emotion, pain and duress that anyone can imagine, and worse. Only He purposely allowed the suffering to be inflicted – for our benefit.

Jesus let Himself be put down, because before Him, no one could get up again. It's a little like a boat, full of non-swimmers, that capsizes. The rescuers that jump in to save them are deliberately putting themselves in the same dangerous, shark infested waters, yet with great distinction. The difference is the rescuers can not only swim, they have the strength, skill, heart and equipment to save others. (Not to mention the tools to repel the sharks.)

Jesus is our rescuer. He came in to the same cruel world where we were suffering and helpless, putting himself in harm's way to save any that will take his strong, extended hand.

Jesus did die for us. He did. But He did not stay dead, that would have been incomplete. He not only defeated the sin that caused us to be trapped, He defeated death. Death could not hold Him.

Because of His victory, we likewise can defeat what has 'killed' our souls. We can come back to 'life'. Our souls can be resurrected. Our true personality. Our real identity. Our true selves. *"Jesus said unto her, I am the resurrection, and the life: he that believeth in me, though he were dead, yet shall he live:..." (John 11:25)* Imagine that, alive- - really alive again! That's what Jesus made possible by the sacrifice He gave.

Jesus so willingly offers this liberty to us, all we need to do is take it. Let me share with you some initial steps to get you going. First, today, at some point, you need to decide that this is the last day you will suffer from this.

Secondly, read over some of the scriptures that have been shared in this book. Choose to believe that God intends the same blessings for you as anyone else, after all, He has no favorites and He doesn't change. Next, you need to ask the Lord to help you, heal you and make you over again inside. God alone can do this.

Finally, choose to accept His kindness and enjoy it. It is okay to smile and be happy. You can dress becomingly if you like. See yourself whole again because you are, despite your memories, for they do not define who you are.

You and I were born to be worshippers of the true and living, all wise God. Worship is the way to totally forget about yourself and come out better. Good deeds might bless others, but when you worship God, you bless Him. That alone will bless you as a person. Think about the man from Gadara that was miserably possessed. No matter how fierce his torment was from that legion of devils, not one of them could stop him from worshipping God.

I cannot describe to you the peace and total fulfillment one comes to experience from drawing close to God. So get close to Him. Worship Him by intently and reverently reading His Words. Worship Him with your talents. Worship Him with your kindness to others. How do we know that Jesus did not go to the man because he knew that the man had a heart to worship in spite of his trouble?

That man could have cursed at Jesus and blamed Him for not coming sooner. But he was such a worshipper that in fact, the man didn't even ask to be delivered. He simply worshipped the living God. When he worshipped the One called 'Wonderful', 'Counselor', 'Prince of Peace' and 'Mighty God', then the Lord showed every aspect of His names to this worshipper, and delivered him.

"The Lord is nigh unto all them that call upon him, to all that call upon him in truth. He will fulfil the desire of them that fear him: he also will hear their cry, and will save them. The Lord preserveth all them that love him..." *Psalm 145: 18-20*

We should never allow ourselves to see God only as the one to ask for help, although He is. We must begin to see Him for who He is. He is God. Our respect, reverence, commitment, devotion, allegiance and honor to God should take precedence over every thing else that we do. The Lord is the only true and living God. He is not a sounding board to air our grievances or one to cast blame for our ills. He is the Lord whether everything in our lives is sunny and bright or if we have times that seem dark and gloomy.

"Wilt thou disannul my judgment? Wilt thou condemn me, that thou mayest be righteous?" *Job 40:8*

I don't know of anyone who lived in such a dire situation as the man that lived in the tombs, but I do know how fascinating and humbling it is to see a true worshipper of God engulfed in praise. It has to be one of the most beautiful sights to behold. This is how I want to be, how we all should want to be, living to worship the one true God.

Life if too short; don't be stuck in yesterday. A new day has arrived. No more are you a victim but a valiant overcomer. Your emotional prison no longer has the capability of holding you, for the strength of your new faith in the Lord has caused you to outgrow the shackles that had you bound.

This is a whole new you now. You are not who you were before you were ravaged, but you are better in spite of it. Now you know where to go to be protected, respected and accepted. Now you understand and know your enemy- - and you also know that he is incapable to conquer you because you belong to God.

This is a great time because your soul is restored, your spirit is refreshed and you have a renewed respect for your body. Now you know that God is your comforter. He can and will avenge you, deliver you, heal you and show you a whole new way to look at how much he values you. See, today- - you understand.

Today you understand and know that God <u>was</u> really there all along. He saw that whole horrible attack on your soul and His heart. But you also fully know that He is the one who made you and He alone can and will make you over inside. This is a great- great- great day.

You are no longer bound with cords of fear, bands of shame, ropes of anger and shackles of hate. Jesus has freed you and indeed you are free. No one can take you away from Him, even if they tried.

Christ's love for you is greater than any difficulty we will ever face. No pain is too painful. No horror too horrible. No attack too devastating. It was all a distraction, a set-up and a trap. But it didn't work- -you are still alive and well, ready to face the day being who you were born to be. Not ruined by a remorseless wretch with no hope, but restored by the marvelous God who loves you enough not to let you stay ravaged, relentlessly.

When you completely give this over to Christ, your pain, shame, loss- it will be over. You will be new. Different. People will expect you to be the victim –the one who had been hurt so bad. They will be coming to you, too –late, ready to somehow make the situation better. They will attempt to put their own versions of consolation on a very dead part of you. But just like Jesus, you won't be there anymore. You will rise again.

"And he saith unto them, Be not affrighted: ye seek Jesus of Nazareth, which was crucified: he is risen; he is not here: behold the place where they laid him..." *Mark 16:6*

Try to understand, this bombardment of emotion is not a simple natural part of your 'life-as–you-are–now-that-you've-been-victimized' this is intentional, emotional warfare.

Satan is called the enemy of our souls for a reason. Your soul is the part of you that must make decisions. It is where you determine the kind of person you will be and how you will affect the world around you. It is where your conscience sits and where your heart aligns to embrace the deeper issues of life.

See, the enemy's tools are not one but many, for one tactic alone could not as easily turn you against the Lord's will for your life. An arsenal of weapons was released upon your mind through the battering ram of sexual assault. Hate, fear, anxiety, bitterness, lust, pride, shame, guilt, sorrow, grief, unforgiveness---all alighted on you at the same time, and you can pinpoint when that time was.

But thanks be unto God!!! No weapon formed against you shall prosper! Regardless of all that the enemy attacked you with, none of his artillery can destroy the real you. God is standing too close to you for that to happen and that is why you are still breathing. As long as you are still breathing, you can make it. God has continued to allow you to be here on this earth. He alone can help you to accomplish what you are here to do.

This is how you rise up as Christ did, all you have to do is trust Him with your life. He left detailed instructions for us to live by, there is no reason to stay in the dark. Trust the One who is risen and you, too, will be resurrected.

Yes, but…

During the attack, God was very present.
Still, I once asked God, were you there when
I was made a victim?

Then I thought of the hymn, 'Were you there when they crucified my Lord?'
Suddenly I understood.
The Lord showed me a different vision of what happened.

When I was being ravaged, Jesus was right beside me only
He was on a cross, right at my feet, yes…
but not as a victim.

When the attack was over, I died inside.
Jesus died too, yes and they laid his body in the grave…
assumedly as a victim.

On the cross He made a way for me to be forgiven and to have the power to
forgive anyone who had hurt me, even the assailant…
who in a hidden way, yes, was too, a victim.

My mind went into a dark pitiful grave of a place I can't describe.
Jesus went into hell literally, yes,
but not as a victim.

In the grave Jesus made it possible for me to have the power to be delivered and
restored
even from what had killed my very soul and yes, the soul of
many other victims.

Jesus rose again victorious. When He arose, He made the way for me to start
over and
begin life anew. It is now my turn to rise, and so yes, I shall,
but not as a victim.

Journal Entry About God's Deliverance:

Family & Friends

This section was very key as it focused on deliverance-which is something that must take place in every victim.

The man who was possessed by demons for (Lord knows how long), suffered in misery day and night. When Jesus came in answer to his cry, He made him completely well. People had previously tied him and tried to tame him but nothing worked.

Just the same, we cannot expect a victim of rape to respond to our methods of easing their pain. We may want them to stop crying and just 'get over it', but that only binds them worse as they then have the added pressure of trying to please us by pretending to be okay.

What they need is a divine intervention from the Lord- - a miracle.

Once fear has entered in, a victim may think differently about everything. They make decisions based on the tormenting thoughts that fear constantly produces in their minds. Your prayers, words of encouragement and gentle prompting will help to restore their confidence. Sometimes, all anyone of us needs is someone to say to us –

"You can do it."

So try your best to motivate the one that you are praying for. Believe God for their deliverance, but avail yourself to be a catalytic part of the recovery process. Remember Simon the Cyrenian? He carried Jesus' cross after him and I am quite sure God blessed him as a result. Likewise, you are involved in the journey your loved one is now on and will be blessed for your support as well.

Because you are so close to him or her, you can take note of the emotional changes in mood. You'll be able to gauge when to

speak or when to just extend your arms for an embrace. Pray for guidance in all of this, the Lord will give you wisdom that is invaluable in this healing process.

Sometimes, all you need is to suggest- even insist on an activity that will bring about a better situation to mind. If you haven't done so, you need to begin to keep the Word of God posted. Have scriptures on the walls and over doorways. Keep His word prevalent and on display.

Prepare for those moments when peace has to literally be pursued. (Psalm 34:14). Set the atmosphere to be one whereas in any room, at any time, anyone can have an actual answer right in front of them.

Artworks of peaceful nature scenes, fresh flowers, etc. are very calming as well as nice colors in the room. Just imagine how beneficial they will be in addition to the passages from God's Word written and in plain sight.

Just do whatever you can to help. Your presence is so vital. You remind the victim, by your gentle and kind manner, of the gentle God that we rely on and trust.

That man that we learned about in scripture that was demon-possessed lived among the dead. But he wasn't really living. He was bound in a constant state of torture that was unbearable. Jesus resurrected him. He can do the same for your precious one. Brighten up, don't allow him or her to think that they are living among the dead.

section six:

"...to heal the brokenhearted..."

"And it came to pass after this, that Absalom the son of David had a fair sister, whose name was Tamar; and Amnon the son of David loved her. And Amnon was so vexed that he fell sick for his sister Tamar; for she was a virgin; and Amnon thought it hard for him to do any thing to her. But Amnon had a friend, whose name was Jonadab, the son of Shimeah David's brother: and Jonadab was a very subtil man. And he said unto him, Why art thou, being the king's son, lean from day to day? Wilt thou not tell me? And Amnon said unto him, I love Tamar, my brother Absalom's sister. And Jonadab said unto him, Lay thee down on thy bed, and make thyself sick: and when thy father cometh to see thee, say unto him, I pray thee, let my sister Tamar come, and give me meat, and dress the meat in my sight, that I may see it, and eat it at her hand. So Amnon lay down, and made himself sick: and when the king was come to see him, Amnon said unto the king, I pray thee, let Tamar my sister come, and make me a couple of cakes in my sight, that I may eat at her hand. Then David sent home to Tamar, saying, Go now to thy brother Amnon's house, and dress him meat. So Tamar went to her brother Amnon's house; and he was laid down. And she took flour, and kneaded it, and made cakes in his sight, and did bake the cakes. And she took a pan, and poured them out before him; but he refused to eat. And Amnon said, Have out all men from me. And they went out every man from him. And Amnon said unto Tamar, Bring the meat into the chamber, that I may eat of thine hand. And Tamar took the cakes which she had made, and brought them into the chamber to Amnon her brother. And when she had brought them unto him to eat, he took hold of her, and said unto her, come lie with me, my sister. And she answered him, Nay, my brother, do not force me; for no such thing ought to be done in Israel: do not thou this folly. And I, whither shall I cause my shame to go?

And as for thee, thou shalt be as one of the fools in Israel. Now therefore, I pray thee, speak unto the king; for he will not withhold me from thee. Howbeit he would not hearken unto her voice: but being stronger than she, forced her, and lay with her. Then Amnon hated her exceedingly; so that the hatred wherewith he hated her was greater than the love wherewith he had loved her. And Amnon said unto her, Arise, be gone. And she said unto him, There is no cause: this evil of sending me away is greater than the other that thou didst unto me. But he would not hearken unto her.

Then he called his servant that ministered unto him, and said, Put now this woman out from me, and bolt the door after her. And she had a garment of divers colours upon her: for with such robes were the king's daughters that were virgins apparelled. Then his servant brought her out, and bolted the door after her. And Tamar put ashes on her head, and rent her garment of divers colours that was on her, and laid her hand on her head, and went on crying. And Absalom her brother said unto her, Hath Amnon thy brother been with thee? But hold now thy peace, my sister: he is thy brother; regard not this thing. So Tamar remained desolate in her brother Absalom's house. But when King David heard of all these things, he was very wroth. And Absalom spake unto his brother Amnon neither good nor bad: for Absalom hated Amnon, because he had forced his sister Tamar.

And it came to pass after two full years, that Absalom had sheepshearers in Baalhazor, which is beside Ephraim: and Absalom invited all the king's sons. And Absalom came to the king, and said, Behold now, thy servant hath sheepshearers; let the king, I beseech thee, and his servants go with thy servant. And the king said to Absalom, Nay, my son, let us not all now go, lest we be chargeable unto thee. And he pressed him: howbeit he would not go, but blessed him. Then said Absalom, If not, I pray thee, let my brother Amnon go with us. And the king said unto him, Why should he go with thee? But Absalom pressed him, that he let Amnon and all the king's sons go with him. Now Absalom had commanded his servants, saying, Mark ye now when Amnon's heart is merry with wine, and when I say unto you, Smite Amnon; then kill him, fear not: have not I commanded you? Be courageous, and be valiant. And the servants of Absalom did unto Amnon as Absalom had commanded..."

2 Samuel 13:1-30

Chapter Sixteen: God Hears Our Silence

"But hold now thy peace, my sister..."

When Tamar's brother spoke this phrase to his sister, we can tell that it was full of passion and consolation. He was letting her know that he would take care of her. He would handle the whole situation. He immediately acted on his word by bringing Tamar into his home to be safe. Safe from any further harm by Amnon. Safe from the stares and glares of the public who would most certainly notice that she was no longer wearing that special garment worn by virgin princesses.

When Absalom said, *"Hold now thy peace"*; it was more than just a statement for her to keep quiet--but something deeper. His statement to her was an admonition to hold on to her peace. Or yet still, to not allow what happened to disturb her spirit to the point where she was no longer herself.

Absalom wanted Tamar to be herself; the same princess who was an honorable and wise woman. The true lady who was free to go about doing good for others. One who was wise enough to know the ways of her people as well as smart enough to know how to handle the toughest of situations.

The very same one who was not too arrogant in her position as the daughter of the king to do a menial task for someone in need, like cooking a meal. The peaceful, free woman who was kind, comfortable in her own skin, fearless and wise, is the woman Absalom was entreating to. He spoke to the Tamar inside the wounded sister he then beheld.

"Remember them that are in bonds, as bound with them; and them which suffer adversity, as being yourselves also in the body."

Romans 15:1

Absalom looked past his sibling's torn garment, past the tears in her eyes and beyond the anguish in her voice. He sought out instead the strong, prudent daughter of the anointed king, that he knew and loved. Absalom didn't want his sister to be anything less. Unfortunately, though he could somewhat protect her, he could not restore and heal her.

Absalom avenged Tamar's honor a couple of years later by killing Amnon, but his actions only made things worse. Tamar spent the rest of her life in Absalom's home, and the Bible records that there she remained desolate. The word desolate is defined as follows:

1. devoid of inhabitants and visitors: DESERTED; 2. Joyless, disconsolate, and sorrowful through as if through separation from a loved one; 3. Showing effects of abandonment and neglect: DILAPIDATED b: BARREN, LIFELESS c: devoid of warmth, comfort, or hope: GLOOMY.

Absalom, try as he may, could not do what Tamar desperately needed. His anger, vengeance, protection and hospitality only had so much worth. Tamar needed a new garment. Not one of a virgin, but one of a woman of active virtue. A woman is virtuous only from her heart, for the spirit is where virtue is developed.

"...to appoint unto them that mourn in Zion, to give unto them beauty for ashes, the oil of joy for mourning, the garment of praise for the spirit of heaviness; that they might be called trees of righteousness, the planting of the Lord, that he might be glorified..."

Isaiah 61: 3

True, Absalom tried, but ultimately, his best efforts could never have satiated her soul. What she needed was a touch from God. Once the Lord intervenes in a situation, there is healing, there is life, there is newness. Sadly, Tamar didn't seek the Lord. In fact, she was never quoted again in scripture. Was she silent for the rest of her days? Some wounded souls do turn to a shelter of silence. They shut the world out, not allowing anyone to get anywhere near them.

I imagine that choosing to be mute brings some a feeling of safety and a certain level of control, afterall, no one can take *that* from them. Silence is, for them, a place where they can dwell without being detected. An invisible cloak of protection from being emotionally exposed.

I could go on and on of all the analogies that could explain the state of silence that pain produces, but I won't. There is no satiable reason to become willfully mute, for though silence may be an escape to some degree, it can also be a perpetual prison.

"The words of a man's mouth are as deep waters, and the wellspring of wisdom as a flowing brook." *Proverbs 18:4*

Speech is one of our greatest means of communicating who we are inside and what we desire. It must be made clear, however, that I am not addressing a mere sharing of an encounter, but the silencing of one's true self. What happened to that person who once loved to dance? Where is that guy that was everybody's buddy? Why can't we hear that familiar sound of singing that used to be a constant pleasure?

Whatever part of the real person, or more personally, the real you, that has become desolate has to be revived. Actually, a better word would be resurrected. That part of you that has been buried needs to be unearthed out of that pit of despair. My book, "No Praise in the Grave", offers further encouragement on just how to come away from that state of emotional deadness that many are stuck in.

You must see that the simple pleasures of life that you once enjoyed are still there for you. Don't allow the horror of what happened to you rob you of your future joys as well. There are sunsets that will astound you that you have yet to see. Songs, still unwritten, that you will hum for years to come. There are people down the road of your destiny that will completely fascinate you. If you will just imagine with me a little more, you will see that your very presence on this earth is reason enough that there are even those who will meet you and be blessed because of you.

Silence, brought on by trauma, cannot be the last defining event of your life. Regardless of how comfortable, content or safe you may feel by shutting out the world, your heart was never meant to be shut in. It's a little like a rose, which needs several hours of full sunlight each day in order to blossom.

"A brother offended is harder to be won than a strong city: and their contentions are like the bars of a castle. A man's belly shall be satisfied with the fruit of his mouth; and with the increase of his lips shall he be filled. Death and life are in the power of the tongue: and they that love it shall eat the fruit thereof." *Proverbs 18:19-21*

Have you ever noticed that your heart has a lot to say, even when you don't open up to others? Think about it. If you were to pen your heart's full expression, you would write some pretty great songs, poems or memoirs. Maybe a painting, a dance or some other masterful work of art. Our hearts have a lot to offer, we just have to find a way to open up. The most breathtaking contributions to the world we live in have come from those who have produced their work from the heart.

Even in our silence the Lord hears us. He knows the heart of every man. You must ask Him to make your heart free again. Absalom didn't know how to repair his sister's heart. Judging by his actions, it's safe to say that if he could have healed her, he would have.

Jesus, on the other hand, is far greater than any brother could ever be. He knows our deepest part-good or bad, and He alone can make us whole again. He hears our silence.

When the rest of the world is trying to figure us out, He already understands who we really are. We never have to explain ourselves to

Him. Jesus knows our actions as well as our motives. While others are doing their best to help us, and often failing at it, Jesus knows exactly what we need before we ask. He stands ready to help us.

The great thing is, His solution makes things better, not worse. His actions won't bring grief or regret to anyone. His counsel is sound and His answer is irrefutable. There is no reason to suffer in an empty place of no hope, not when Jesus is the Savior. That's why He is called Wonderful, Counselor, Mighty God and Prince of Peace. The Lord God is the only one who is all we need.

Call Him by name, any of His names. He will hear you-even if no one else does, and He will respond to you with the answer your heart requires. For anyone who has ever whispered to the Lord: -"please, just help me to feel a little bit better-take this pain away even for just a little while"—God hears you. So, "hold now, your peace."

"Unto thee, O Lord, do I lift up my soul. O my God, I trust in thee: let me not be ashamed, let not mine enemies triumph over me. Yea, let none that wait on thee be ashamed which transgress without cause. Shew me thy ways, O Lord; teach me thy paths. Lead me in thy truth, and teach me: for thou art the God of my salvation; on thee do I wait all day. Remember, O Lord, thy tender mercies and thy lovingkindnesses; for they have been ever of old..."

Psalm 25:1-6

Chapter Seventeen: The Vulnerability of Virtue

One of the reasons the attacker wants to defile you is because you carry within yourself a beauty he does not possess. In him is guile, filth and evil. When he sees in you goodness, trust, and innocence—it angers him and he wants to take from you what he feels he cannot have- -what he thinks no one else should have either.

There is beauty- true graceful, resplendent beauty in what is holy. What is pure. Untouched. A rapist lives with his own degree of darkness and it irritates him to see the light of someone else's joy. He wants something he feels he cannot have. If he doesn't have it, then to him no one else should. In fact, he wants to punish anyone who dares think that he or she can just live such a carefree life when he exists in cold misery. He feels punished and shortchanged. So he turns into what is in him. It wasn't you that he was after, it was what you possessed.

Is there a difference between a virgin and a woman of virtue or are they one and the same? Herein is the perfect mystery. One not founded in stone but carried out throughout time the world over. It is an unspoken matter that demands a certain reverence and contemplation. What is the answer?

To say that a woman who is a virgin is virtuous may lend to her credit that she does not merit. Why? Simply because virtue primarily is a matter of character. Of course, if a woman is not a virgin, no matter how well behaved she is, can she really be regarded as virtuous?

The two descriptions we are dealing with now should, naturally go hand in hand. Often, as is commonly known, they do not. Where is the balance? We cannot punish all unmarried women that are not virgins; To do so gives no consideration for those who have lost their virginity through no fault of their own.

Am I implying that the ones who willingly gave their virginity away should be cast aside? Afterall, it is a very common practice to do so. If I were unwise, I would. However, because I have learned, largely through my own experience, of the grace of God, I am eternally certain that God treasures our hearts.

A virgin holds a certain birthright and promise of possessing at least one expanse that is untouched by any man. Because so much of the world is tainted and perverted, to have something unspoiled is a beauty that is highly prized and often vehemently pursued.

As precious as virginity is, one can never foolishly believe that all will respect it, but on the contrary, there are many who aim to defile it. Even for the very sake that it is sacred. Virginity is not nor can it ever be merely a physical attribute. It is spiritual.

Here is where we differ from the animals. The animals and plants, if for no other reason, diligently reproduce to keep the cycle of life going as the Lord commanded them when He created them. In many ways we can see various species of the wild kingdom behave in ways that seem to mirror our own. At times we see courtship, violent displays of dominance, self-sacrifice, homemaking and providing for the young. Contrary to many scientific theories, man is not an animal. Man is a spirit that possesses a soul and lives in a temporary body.

No one can touch the real you – the you that is spirit unless you allow them. They may commit crimes against your body and speak offenses to your soul, but your spirit? Not possible. You have to allow the offense to settle inside of you…you do that, not the offender- - he doesn't have that kind of power. No one, not by might, nor by permission, can ever take away the spirit of another. Not even by sexual crime.

Tamar was raped by someone she trusted. Of all of the individuals we have studied so far that were sexually assaulted, she was the only one that suffered directly because of a sibling. Afterwards, she allowed her anguish and offense to dictate the rest of her life. It was almost as if she believed that her virginity was who she was. True enough, being a virgin is special and important, but it is not one's only identity in and of itself. It is not the virginity that is so valuable, but the virgin- that is, the *person*. Let me put it in simpler terms.

Before any of us could talk, we were all mutes, so to speak, when we were infants. Did we not still have voices and were able to communicate our desires? Before we could crawl or walk, we were helpless to a point, but were we paralyzed? Did we not have energy and movement? Before we could read or write, were we without intelligence? Of course not.

What we possess and are able to do does not give our lives meaning. For all of our skill, talent and possessions can be lost to us at any time, but our worth will never diminish.

All virtuous women are not virgins any more than all virgins are virtuous. There is no mystery in that. No one else can even know that a woman is a virgin without a medical examination. Even the 'blood' that is expected is not necessarily a certainty of proof. A gynecologist once told me that the only time blood is shed is usually if the man is too rough with the virgin. The hymen should be pushed pass gently, and cooperatively, not broken violently. The act is a sacred, silent and secret embrace meant to consummately seal a marriage covenant, changing the male and female, each, into a spouse. Virginity is to be respected and kept in its perspective place. Outside of that, idolatry and arrogance can turn it into more than what it is intended to be, which can be dangerous. Celebrating virginity is one thing, idolizing it is quite certainly another.

Virtue, just the same, should be respected in the right perspective. Otherwise, one can make an idol of someone known to be virtuous, and worship them instead of the glorious Lord who taught them to be that way. The similarity of virginity and virtue is this – they both are of God.

The Lord gives virginity to everyone born in this world. In a way it symbolizes a promise of the greatest kind of love. Love that willingly gives what is most sacred, most guarded, most honest.

To marry someone perpetuates an exchange of the gift of virginity (sacred love) over and over again, for the heart validates it. This is why a person that is not a virgin can marry and, by the restorative grace of God, enjoy a fulfilling marriage with no shame or guilt.

I myself know how virtueless a virgin can behave. Heaven is a witness to the shameful things that many have done in secret, even as virgins. The Lord wants each of us to walk in virtue, whether we are virgins or not. Jesus said (in essence) that if you lust in your heart after someone, you have already committed sexual sin with them. That goes for virgins as well. (See Matthew 5:28)

When it comes to virginity and virtue, the best thing you can have is not one or the other, but a respect and understanding of both. For what matters above all is that we honor God. Never have other gods before Him, that is the very first commandment, remember? (See Exodus 20:3) Virginity or virtue, neither one should be such a 'god' to you that you forget that you were created to please the true God.

Tamar allowed her life to emotionally end after the rape. God never intended her to be desolate that way. She was created and put on this earth for a reason and that rapist didn't have any power to change that. Tamar chose to let the loss of her virginity become the reason she refused to honor God by fulfilling her destiny. But I don't condemn her; rape is a terrible thing and without the Lord it is almost impossible to recover from it. The great thing about Tamar's account is that God worked it out for the good anyway, because so many can learn from her.

Her brother may have stolen her physical virginity, but she destroyed her own virtue. Virtue is active and meaningful. It is productive, industrious and helpful. Virtue is honorable and needed. It makes families peaceful and communities strong. It makes peace between nations and displays the glorious characteristics of God. Virtue is meant to be shared with the world, virginity is not as it can only be shared with one. See the difference yet?

Virtue is kind and loving, faithful and honest. Patient, full of hope and a force to reckon with. It is a choice. Virginity is unique in that it is a God-given birthright. Understand the difference and consider them respectively. If you confuse one for the other, you can become as Tamar – unfulfilled.

So now that you know, take back what is rightfully yours… the beauty of God's holiness. You can do so simply by living your life consecrated to God. Totally separated and dedicated to His will. You don't have to become a dirty person just because you have been soiled. The Lord can cleanse away every sin and shame.

"…Behold, the Lamb of God, which taketh away the sin of the world." *John 1:29*

So if you believe, you don't have to be an unholy, promiscuous, homosexual, distant, indifferent, bitter and even hateful person because of what happened to you. To do so means you only survived the assault. You can be a loving, tender, sensual and exciting spouse if you will, but you have to let the Lord make you free. Ask Him. God can take the barrier of hurt out of your marriage bed so that your spouse doesn't have to suffer because of what happened to you. Don't let the rapist have that too. Enjoy your marriage bed!

Before God, it is holy. *"Marriage is honourable in all, and the bed undefiled..."* *Hebrews 13:4*

For those of you who are not married, you don't have to add insult to injury by allowing the enemy to oppress you into behaving in a demeaning, disrespectful and disgraceful manner. That is unbecoming of a true human and will only further the loss your of dignity. Do better than that. Don't merely survive it, conquer it. Holiness is key. The rapist wanted to defile you to make you dark like him. But thank God- you don't have to be. Walk in God's light. Follow His ways.

The bible clearly states that sexual sins are against one's own body. (See I Corinthians 6:18) As tempting as it may be, don't do it. It won't make you feel better in the end. You must understand that those desires are not who you are, you must not act on them even on a bad day. *"...Wherefore take unto you the whole armour of God, that ye may be able to withstand in the evil day..."* Ephesians 6:13. Before you were attacked, you were different. Many of the feelings you have are a direct result of what happened to you, and would probably have never otherwise crossed your mind.

When a person is violated, defiled and stained sexually, the violator leaves a trace of his behavior or mindset with the victim. A mark on the soul, if you will. Because of this some that have been offended sexually go on to become sexual offenders. Directly or indirectly, outwardly or inwardly, the offense is more like a leprosy than anything else.

By leprosy I mean it either causes a degeneration of the soul, causing one to repeat the offense on others, or gravitate to a related sexual offense. This may include voyeurism, exhibitionism, masturbation and promiscuity. These sins at times include others and always hurt the individuals themselves. How? Because such sinful acts defile. The question has arisen many times, "If masturbation is not in the bible, how is it a sin?"

Actually, it is in the bible as it can easily be grouped in terms such as lasciviousness, uncleanness, lewdness and the like. Whenever you see these terms in scripture, this is the kind of behavior that it denotes. The reason words like masturbate are not specifically there is because a dark act such as masturbation changes in name from generation to generation. However, the term 'uncleanness' is clearly understood.

How do you know it is an offense? You can measure it by asking yourself whether or not it is an act that pleases God, who is holy.

"...Who being past feeling have given themselves over unto lasciviousness, to work all uncleanness with greediness."

Ephesians 4:19

Some may argue that it is only natural. Then consider this, the masturbating act itself, if a stranger initiated it, would cause emotional trauma and duress. If an armed and dangerous stranger demanded you to commit a lewd sexual act in his presence to yourself, you would be horrified. It would scar your psyche and soil your spirit. If the act is this offensive then how can it be any less offensive when you do it before God- who sees and knows all? It mocks God's perfect ways and causes the heart to grow callous. Because it isn't against the laws of civilization, we tolerate this as if it is okay, even laughable. But there isn't anything funny about being abominable before God.

Haven't we learned so far in this study that sexual sin is never satisfied? For one to continue to practice such opens them up to accept, tolerate or even consider worse behavior. If we were to ask any sexual offender if he or she ever masturbated, indulged in pornography or any other secret sexual sin before they bothered anyone, what do you think their answer would be? This is no new information.

"Because sentence against an evil work is not executed speedily, therefore the heart of the sons of men is fully set in them to do evil." *Ecclesiastes 8:11*

If you weren't doing such things before you were defiled, you don't have to carry the leprous disease of that offender and act as vile as he did after the fact. Through Christ, you can overcome. Nothing is impossible with God, and besides this, He is the only one that can heal leprosy, natural or emotional, anyway.

The enemy of our souls wants us not to care about our bodies and perform all sorts of vile, shameful acts. He knows that God intends our bodies to be holy temples for His very presence to dwell in, a physical witness of His reigning goodness here on earth.

The man that raped Tamar lived in misery before he ultimately forced himself on her. Interestingly enough, she, in turn, lived in misery afterwards. She contracted it from him just like a disease. This does not have to be your story. Your virtue is too valuable for you to let it be vulnerable to an outward act of offense. Appreciate Tamar's story and learn from it by not giving up the way she did. I know its hard, believe me, I know, but you must refuse to be any less than what you were born to be. Don't carry yourself as one that is defiled, damaged or weak. Resist the natural inclination to hide yourself away from the world. You are too precious. Walk in your virtue and bless the world with your wonderful spirit- - unmoved, undaunted and certainly not desolate. God can give you a brand new life, but He will never force you to accept Him. Forcing is not God's way.

"Jesus answered and said unto her, Whosoever drinketh of this water shall thirst again: But whosoever drinketh of the water that I shall give him shall never thirst; but the water that I shall give him shall be in him a well of water springing up into everlasting life."

John 4:13-14

Chapter Eighteen: "Were You There When…?"

It took me thirteen years to verbally ask the Lord the question of which I already knew the answer. I never dared allow myself such liberty before. I thought it disrespectful and irreverent to question God and ask His whereabouts during one of the most traumatic times of my life. So for well over a decade I held to that pining unanswered notion. Although I knew in my heart that God was there with me, my soul needed to hear God say it. I wanted God to speak to me and give me the affirmation I desperately needed to have.

For some reason, it seems that peace is the very first loss that occurs after a heinous crime. For some, a strange sound or unfamiliar voice took away the peace. For others, a sense of something unusual or that certain facial expression you saw just before your life was drastically altered.

Once your peace was gone, it was gone. Gone as if it were completely destroyed. Suddenly there was no peace in the normal comings and goings of a day. No peace in sleeping, or in waking up to face the day. No peace in traveling or meeting new people. Or being alone too long or too far from someone within close emotional earshot. The peace to dress well, style your hair again or wear perfume just because you enjoy it.

How do you get the peace to just live? Where is it? Is it obtainable? If so, how? That is the first thing I believe God wants to restore to you and those who have been hurt this way. Peace.

One of the greatest aspects of the heart of a Christian is the ability to forgive others. Because of what Christ Jesus did on the cross, every believer has in this regard, the same capacity to forgive as He. Jesus made a great gift available to us all when He voluntarily gave His life. But that's not all He gave. In (Luke 23: 34) Jesus prayed "Father, forgive them; for they know not what they do." Here was the one individual who was treated more unjustly than anyone else ever was, the very Son of God. He never did anything wrong, not even being selfish. Yet Jesus forgave those who beat, mocked, humiliated, lied on and crucified Him.

We, as partakers of His divine nature can and must do the same as He. Jesus Christ gave his blood to save us spiritually, just as we donate ours to save a life in the natural. From this holy transfusion, we have new life coursing through our spirits. The dis-ease is removed because of the sacrifice that was made by the Christ. That means we are able to forgive those who trespass against us as He forgave. Jesus words are very clear about forgiveness.

"And when ye stand praying, forgive, if ye have ought against any: that your Father also which is in heaven may forgive you your trespasses. But if ye do not forgive, neither will your Father which is in heaven forgive your trespasses." *Mark 11: 25-26*

"Then came Peter to him, and said, Lord, how oft shall my brother sin against me, and I forgive him? till seven times? Jesus

saith unto him, I say not unto thee, Until seven times: but, until seventy times seven." *Matthew 18:21-22*

Basically, we must take time during our normal praying to forgive those who have harmed us. That means during our prayer time when we give God our all, pouring out our hearts to Him. We must go to God right away to get our hearts healed and our souls restored. It's not the time to hold back. Moreover, we must learn the importance of immediately forgiving, not allowing our anger to get the best of us. We cannot let our offense go unchecked.

I want to focus on the message that Jesus gave us in the 70x7 lesson. It is not a hypothetical parable of sorts. If you approach it literally, you will find that you have defeated the attack of the enemy of your soul single-handedly with one weapon- - forgiveness.

When someone harms you, depending on the severity of the offense, you will undoubtedly replay the incident in your mind over and over. If you were to record the number of times you thought about, relived and repeated all the details of the occurrence, it will most likely be quite numerable. The harmful thing about this is that each time you do, it tends to hurt all over again. You feel anger, animosity, and even hatred towards that person or persons. Meanwhile, they feel no pain.

For many of us, just imagining someone harming us 70x7 times a day is too much to fathom. For others it is a painful reality, for they live with those who hurt them regularly.

In either case, the secret to freedom is found in the 70x7 principal. Forgive each time you are hurt, then forgive each time you recall it. Most likely, you will remember the incident many, many more times than you actually experience it. It's understandable that many people 'block out', or refuse to think about what happened to them.

But denial or burying something (so to speak) inside our hearts does not bring about liberty from the pain of it all. Forgiveness does. We must practice Christ's law of forgiveness. With that we have the assurance that when we forgive, we will be forgiven. If you do not forgive, you will not be forgiven by God of the sins you committed against Him. At some point in our lives, we all need God's forgiveness.

"...Repent ye therefore, and be converted, that your sins may be blotted out, when the times of refreshing shall come from the presence of the Lord..." *Acts 3:19*

"...And the times of this ignorance God winked at; but now commandeth all men everywhere to repent..." *Acts 17:30*

See everyone must get in right standing with the Lord at some point in their lives, He commands it. Unforgiveness is one major way to keep that from happening. How can we forgive 70x7 times a day? It is very simple: Every time you think about what happened—forgive. Every time you dream about it, forgive. Every time you are reminded in some way, forgive. Whenever you are asked about it, forgive. Forgive when it hurts. Forgive when you are annoyed. Forgive when it angers you.

Forgive verbally. Forgive silently. Forgive while in prayer to God. Forgive as a testament, to your friends, of obedience. They may not forgive the one who hurt you, but you must, how do you know that your peace won't free them? The more you deliberately practice forgiveness, the more healed you become. The freer you are. The more restored and renewed you become. Forgiveness is a spiritual law of God. It is a key to His kingdom and an entryway to His favor, an invisible eraser of sin's effect.

Forgiveness helps to usher in your own deliverance and helps to maintain your peace. If you ask the Lord to help you to forgive and you determine in your heart that you will not hold the offender's sin against them any longer, you will be free of the offense.

The more often you forgive the more you will be able to truly forget it. You defeat the pain when you forgive. In your obedience, you simultaneously resist the enemy and he flees from you. This breaks down his attack plans and destroys his stronghold on your mind.

Jesus did not tell us to forgive as a suggestion or an option. He told us to forgive so that we will have access to the kingdom and all the love, joy and peace it provides. I do sincerely hope that you believe me when I tell you that the moment comes when an entire day, week, even year will arrive, where the assault doesn't even cross your mind. You won't be open to be re-hurt at the thought of it. Your peace will no longer be disturbed. However, if you choose not to forgive, you can become someone you were never meant to be.

I once heard someone say that when a person is raped they either become cold, promiscuous or homosexual. In many instances this is an obvious conclusion. But the truth is a woman, man or child that is raped become in a way transformed, imprisoned and/or lost. A change occurs, not like a naturalistic metamorphosis that is destined to happen, but more like a permanent scarring after a really horrible burn. An unrecognizable soul emerges from a flame that blazed out of control. Therefore, healing must begin immediately.

Anyone who is hurt this way can sink into a dark cave to seek protection. But a cold, dark, damp, and empty cave is no place for human beings created in the image of God. That is not where you are to dwell, for you are the object of God's love. If you can recognize anything at this point, recognize the fact that you are still God's creation. You may have changed in many ways—mentally, emotionally and so on, but the person you really are is still there. God has not forsaken you. He has not left you. He has not turned His back on you. What He has done, is remained God in *spite* of what has happened to you. He's still on the job, ready to hear your prayers and cries for help. He is still the healer and the mender of broken hearts. That's good to know when everything else in your world has changed.

"I waited patiently for the Lord; and he inclined unto me, and heard my cry. He brought me up also out of an horrible pit, out of miry clay, and set my feet on a rock, and established my goings. And he hath put a new song in my mouth, even praise unto our God: many shall see it, and fear, and shall trust in the Lord."

Psalm 40:1-3

It took me thirteen years to get the nerve up to ask God that one question I thought to be most disrespectful. But out of the abundance of my heart, my mouth did indeed, eventually speak. I didn't ask the Lord why He had forsaken me, but in a round about way, I guess I did. I sat in the office one night thinking about an old hymn I had recently heard again. "Were you there when they crucified my Lord?"... is what the singer earnestly cried. For some reason those words seemed to ring out to me in a very different manner. "Lord, were <u>you</u> there when I was..."; was my question of the hour. Then I just went down the list of everything that happened and everything that I could recall about those awful times in my life when I had been ravaged relentlessly.

Were you there when he did (this)? Were you there when they did (that)? Did you hear when she said I was to blame? Did you notice how they looked at me after they found out? Were you there when I prayed to die, and died anyway, only not physically? Were you there when I could not breathe anymore? Did you mean for my taste-buds to know (that)? Was my heart meant to pound that hard? Lord, were you there? Were my hands meant to be bound? Was I supposed to receive the confusion that was forced inside of me? Lord were you there?

My questions that night were as numerous as my tears, which seemed to have flowed from some locked up reserve. But my answer came for each and every inquiry. God had not left me. He was there the entire time. If He had not been there- - I wouldn't be here. I wouldn't be living and breathing and being who I was born to be. Otherwise, I would be alive, but barely. Merely existing or walking around dead inside. It may have felt that way, but I wasn't dead. God kept me.

If I could give an analogy of God's presence in tough times, I would liken God to being the captain of a great ship. I would say that the ship itself would be God's plan, under His command, of course. The sea would be the destruction caused by sin. As long as we stay in the ship, we are pretty secure and safe. We are going to great places and will one day be on dry land. Sometimes, on the journey, terrible storms arise. Winds, strong undercurrents, crashing waves, torrential rains, lightening and thunder are so much like the problems we face. Problems like sickness, grief, poverty—even sexual crime.

The sea itself is the very magnet that causes most of these storms. If you happen to be sailing high on the deck when one of these storms hit, you can get knocked or blown overboard quite easily. That's what happens when you are hit with a terrible trauma. You are knocked off of that great ship by a fierce tempest into that deep dark sea. If you stay there, you will drown. But if you will just lift up your eyes you will see that the Captain is still there and the great ship has not left you behind to drift all alone. The Captain will make every effort He can to bring you back on board the ship and will not cease until you take his outstretched hand.

"...for He hath said, I will never leave thee, nor forsake. So that we may boldly say, The Lord is my helper, and I will not fear what man shall do unto me." *Hebrews 13:5-6*

See, God has not forsaken you. He is right with you. He saw you when you were literally knocked out of His ultimate plan for you.

God is aware of the storms that occur in this life, but He knows that His ship is sound. He will take you to where you are destined to go, no matter how many times he has to stop along the way to rescue you. Don't stay in those deadly waters. Don't hang on to the debris that floats around to feel safe, reach for the outstretched hand of the Lord instead.

You don't have to become promiscuous, indifferent, coldhearted or abandon your natural affection, that's just debris. When your heart is crushed its easy to turn to such a sad state of being. But sinful practices are only something that keeps you from sinking for a little while, but still leaves you in peril. There is a life preserver cast out to save you, He is the Christ—take hold of Him and He will pull you back where you belong.

Believe me when I say, that " Were you there?" question I asked the Lord, comes to many people when sorrow and pain overtake our lives. It's natural to want to ask Him. So ask Him, but from your heart, not from your anger. The Lord doesn't mind you asking for understanding, and He will tell you all that you need to know when you ask in faith. Don't distance yourself from God for fear of saying something wrong. Call out to Him. Tell Him how you feel and ask Him to explain what His purpose for you is because you need to know. Once you take your heart and soul to the Lord, the healing will follow. The emotional warm blankets and hot soup and steaming bath of comfort come as soon as you are truly back where you belong.

If you can, ask the Lord, even now, to rescue you from this deep place you have been cast into. It isn't where you are meant to be and it is not where God wants you to stay. Just ask Him: "Lord, pick me up, bring me out of this nowhere place, I want to be where you are, not just have you in view. I want to hear your voice in my heart and feel your love in my soul. I need you to warm the coldness of this horrible crime away from me. Lord, help me." God doesn't want your heart to be broken another minute, He wants to make you whole again.

"Blessed are they that mourn, for they shall be comforted."
Matthew 5:4

Tamar didn't know that God was with her anymore than some of us. If the Lord hadn't made a way for me to find out the truth, I imagine I could have ended up as she did or worse. If we stay out in the cold long enough, we will freeze. If you allow your life to stay hurt, as Tamar did, never seeking the Lord for help, the same result can happen. You have to make the choice to trust God.

I have learned that sexual crime, because it hurts a life on so many levels, can really substantiate a persuasive argument that God doesn't care. All we have to do is read His own words to know that this is not the truth. He does care for us. There aren't enough words in my vocabulary to convey to you how deep God's love is for us, and my understanding is limited. We must trust Him without exception. You wouldn't turn away from God because of sickness, why is sexual crime the deal-breaker? Is it *that strong* that it can make you completely turn your back on the One who truly loves you no matter what? No it is not! Sexual crime is only one of <u>many</u> evils that we all endure in this life. It is not, nor can it ever be, more powerful than our destiny. Let us not be mistaken by its depth of pain, believing that it is indicative of God's care, it is not.

We hurt when a loved one dies, yet we still pray. We weep when we get a bad report from the doctor, but we still serve the Lord. When our money is low, still we find a way to worship God and if our material things are suddenly lost, we manage to praise God for sparing our lives. Why, dear one, why, can't we determine in ourselves, that even if the unimaginable, unthinkable and unspeakable happens, that we will yet love the Lord? He is not our enemy!

The assailant, whoever he or she was, may have ravaged your body, but no amount of strength can grant them access to your soul. That place is reserved for God and He will never force himself on you. You get the honor of embracing Him willingly. I am so sorry that this terrible thing happened to us, but please precious one, please...believe me, we cannot let this leave us desolate. We are born to please the one true and living God, no matter what. It is the greatest privilege that there is. Let us not allow anyone, anything, or any tragedy, stop us from the greatest joy of life.

We must continue our quest to journey on to know the Lord who is so sweet. On this earth we will at times have pain, but one day, we won't hurt anymore ever. In fact, God has already made plans to do away with this present evil once and for all. So please, let's endure a little longer, helping others to know the truth as we go. There are too many other hurting souls out there that we can help, we simply cannot waste away in our own pain. Let God heal your storm-weathered heart. Then, tell others that He heals. This is your testimony now...this is part of your mission. Because of Jesus the Christ, we are overcomers, so trust Him, dear one, trust the Lord. You will never be the same again.

"...Therefore if any man be in Christ, he is a new creature: old things are passed away; behold, all things are become new."
2 Corinthians 5:17

"... And God shall wipe away all tears from their eyes; and there shall be no more death, neither sorrow, nor crying, neither shall there be any more pain: for the former things are passed away." *Revelation 21: 4*

"He that overcometh shall inherit all things: and I will be his God, and he shall be my son." *Revelation 21:7*

Journal Entry About God's Healing of Broken Hearts:

Family & Friends

We can see what can happen when a family tries to execute justice on their own. Tamar's brother Absalom turned his life upside-down because of what happened to his sister. Yet for all that he did even to ultimately losing his own life, he could not heal her heart. Nor could he restore her virtue or answer the questions that left her soul un-eased. He only acted in rage, indifference, bitterness and violence. He allowed her misfortune to turn him into a murderer and rebel, overthrowing his own father, who was the king. Tamar wasn't the only victim.

Your loved one wasn't either. Don't remain one. Reach out to the Lord for yourself even while you are seeking answers for her.

You were hurt too.

Don't let that rapist rape you of your joy and peace. You must trust the Lord. Let Him heal you and strengthen you.

Besides, she needs you to be there for her- alive, not on the run to a destructive end as Absalom was. The greatest battle of every victim of rape rages within the victim's own soul. You need to be around to help her get through that. You can't do it if you don't believe that God can restore. Don't lose the faith. Trust in the Lord. He is the Healer -not you.

Finally, my greatest hope is for you to know that God really loves you. All of this work is the result of my seeking the Lord for wisdom concerning the pain that no one wants to speak of. A sexual predator can be male, female, old or even very young. He may be a professional, a relative, a stranger or a person of authority. Who the attacker is, as we have learned, is not the focus, for it was the forces of evil that influenced the assault. The important One to fix our attentions on is the Lord, who heals us, comforts us and sets us free from the mess we were left in. By Him we overcome and become better. By Him, that is, the Lord God Almighty, we can be made whole again.

I'm no authority on any of these matters, but I do know what the Lord has done for me, and I will spend the rest of my life telling the world just how sweet He is. God loves us and will do anything for those who believe in Him with all of their hearts. Nothing can ever change that, not even being ravaged relentlessly.

"Heal me, O Lord, and I shall be healed; save me, and I shall be saved: for thou art my praise." *Jeremiah 17:14*

Before I leave you, I want to be clear on a few more things. If you remember what I wrote about being on the offense as the best defense, specifically, knowing your enemy, then you should not be alarmed to know that he might try to hurt you again either in the same manner or by bringing back old memories. Be aware of yourself, you are no longer the weak, vulnerable person you once were. There has been enough of God's Word in this book to clearly point this truth out to you. Wisdom demands you heed it now, for wisdom is difficult to find later.

Whatever you can do to put yourself in a better position, do so. You may believe that you won't get sick by going out in the cold air without a coat, but why set yourself up? Likewise, you must deny the enemy access to your life, not due to fear, but out of wisdom. Remember the words of the Lord: *"Be sober, be vigilant; because your adversary the devil, as a roaring lion, walketh about, seeking whom he may devour: whom resist stedfast in the faith, knowing that the same afflictions are accomplished in your brethren that are in the world."* *I Peter 5:8*

For the most part, this teaching has addressed a terrible thing that happened to you or your loved one. I will not neglect to remind you that you don't ever have to be victimized again. This is a new day and you are a new person. You must believe in the strength and power of the Lord. Get close to Him and learn to walk with Him, trusting in His merciful, protective hand.

Please understand me, I am not referring to sexual crime alone, but a raping of your spirit by those who seek to overpower you in other ways. You must learn to be strong in the Lord. Refuse to be intimidated by those who are only striving to advance their own cause at your expense. Determine within yourself that you will do what is right in spite of anything else. In relationships, in your family, on your job or wherever you sense that someone is forcefully using their power to pin you underneath their will, turn the situation around in prayer. Be relentless about it. Keep ravaging the kingdom of darkness with your prayers, and you will come to know victory.

"...Submit yourselves therefore to God. Resist the devil, and he will flee from you." *James 4:7*

You were not placed on this earth to satisfy the expectations, lusts and whims of others. You are here by the will of God to cause a positive difference in the world, and you simply cannot do that if someone else has you 'tied' up for their own selfish ambitions. You are not a victim anymore, so forgive them in love, pick up your discarded dreams and live. Through Christ, you can do all things; so do them according to the Lord's direction, not letting anything hold you back ever again. (See Philippians 4:13) If you can believe it, praying a prayer such as the following one can propel you even further into your destiny:

Oh God in Heaven, I am sure that you are the one and only God. I believe with every fiber of my being that you love me beyond my comprehension. That is why I know you didn't hurt me, you couldn't have, you are good and there is no evil in you. So, Lord, my heart yearns for you now. So much has happened in my life, it was easy to get distracted from what really matters and that is getting to know you more and more each day. Therefore, I forgive: (name the individual); for being so cruel and hurting me so deeply. I release them into your hands to change and make them better. Only you know, Lord, how this person came to such an awful state of mind, and only you can transform them into who they are meant to be.

Lord, I want to be changed as well. No longer fearful, angry and crushed, but strong, stable and at peace. Quiet my spirit dear God and ease my mind. Repair all the broken places deep inside of me and show me what I am supposed to do with this life you gave me. Reveal my purpose to me and give me wisdom to walk out my destiny. You placed me here for a reason and there is no reason that I should not fulfill your plan for me. Help me Father, in the name of the your Son Jesus, the one who came to save us. Amen.

Has this message opened your eyes to see the goodness of the Lord? I believe a transformation has already taken place in you either with personal healing or newfound wisdom to assist others that hurt. For those who need further ministry, I would like to suggest our small group workshop: Hiding Place. This is a three- day/ 2 night interactive workshop to help continue the healing work that this book introduced. Just send us a request to be included on the waiting list.

Churches, family, friends and supportive organizations are encouraged to sponsor these dear souls so that they can attend at no cost to them. The scriptures tell us about the four friends of a lame man that not only carried him on a stretcher, they lowered him through the roof of the house where Jesus was teaching, because the crowd would not let them through. Your loved one needs you to carry them to the Savior in the same faith-filled and determined way. Help them to get to this special workshop to be ministered to in person. As with all of SINCERE outreaches, the workshop is of no charge, but the accommodations for this workshop location are.

Our website has more information at www.sincereministries.org Ultimately, our prayer and quest for each one is complete deliverance, total healing and finally…finally…closure.

Other books by Minister Kesha Lee Gore:

"No Praise in the Grave"
A Testimony of Understanding
Published by SINCERE Ministries, Inc.
Copyright ©1999

"Think On These Things"
(31-Day Poetic Devotional)
Published by SINCERE Ministries, Inc.

Copyright ©2000

Resources that were helpful in the preparation of this work:

The New Strong's Exhaustive Concordance of the Bible
ISBN 0-8407-5360-8
Copyright ©1984 by Thomas Nelson Publishers

Dake's Annotated Reference Bible
Copyright © 1963 by Finis Jennings Dake

Webster's New Collegiate Dictionary
Copyright © 1979 by G. & C. Merriam Co.

Thompson Chain-Reference Bible
Copyright ©1964 Dr. Frank Charles Thompson and Mrs. Laura Boughton Thompson

The Amplified Bible
Old Testament copyright ©1965, 1987 by the Zondervan Corporation, The Amplified New Testament copyright © 1958, 1987 by the Lockman Foundation.

Books that may further encourage you:

Believer's Authority
Reverend Kenneth E. Hagin
Copyright © 1985 Rhema Bible Church AKA Kenneth Hagin Ministries, Inc.

Eternal Victim/Eternal Victor
Reverend Donnie McClurkin
Copyright © 2001 by Pneuma Life Publishing, Inc.

Raped
Deborah Roberts
Copyright © 1981 by the Zondervan Cooperation

The Atonement Child
Francine Rivers Copyright © 1997
Tyndale House Publishers, Inc.

Additional Resources:

National Network to End Domestic Violence
5666 Pennsylvania Avenue, S. E. Suite 303
Washington, D.C. 20003 202-543-5566

National Domestic Violence Hotline
1-800-799-7233 (SAFE)

1-800-787-3224 (TTY for the Hearing Impaired)

Prayers for the Readers by the Associates of SINCERE Ministries

Prayer for Spiritual Awakening:

"I pray that you will receive a revelation of the love of God, so that you will never hunger or thirst again."

Minister Denise Coles

Prayer to Receive Salvation:

"I come now understanding my state as a Sinner. I come confessing my sins and accepting that I am a Sinner that indeed needs salvation. I believe in my heart that Jesus Christ died for me so that I may live and live in abundance. Jesus stood in my stead, as the only Savior and way back to my Father God, my Creator. I humbly ask Jesus into my heart to save me and give me a brand new start. I thank God now for my salvation and a new direction, in Jesus' Name, Amen."

Minister Cecilia Hill

Prayer for Courage:

"My prayer is that you would have the courage to embrace the truth of God's word and that you would, in the name of Jesus, I pray, receive this particular promise found in the I Peter 5:10 (Amplified Version):

"*And after you have suffered a little while, the God of all grace [Who imparts all blessing and favor], Who has called you to His [own] eternal glory in Christ Jesus, will Himself complete and make you what you ought to be, establish and ground you securely, and strengthen, and settle you.*'"

Sister Rosie Washington

Prayer for Inner Peace:

"Father God, I thank you that you are my Father and that you are with me at all times. I thank you for your love, your kindness and your faithfulness to me. I know your plan for me is good.

When things seem difficult and I don't understand what is happening around me, I know I can look to you for strength. You have promised, that if I call out to you that you will answer me. That if I draw near to you that you will draw near to me. I thank you for being an ever-present help in times of trouble, and delivering me from the problems that surround me.

Father please give me your peace during the difficult times. Father as I keep my mind on you, you have promised to keep me in perfect peace.

Protect my heart, soul and mind from anything that seeks to destroy me. I thank you again for all you have done and continue to do. In the name of Jesus, Amen."

Deacon Harriet J. Adamson

Prayer for Deliverance:

"I am believing that from these pages you have found healing for your broken heart, God's overcoming power for your times of fear and encouragement for your moments of despair.

My desire is that as you read this book, you were honest, and let your heart be open to the Spirit of God to be delivered. I pray that there has been a true transformation and that it wasn't just head knowledge.

Finally, I'm believing God that not only has this message been transforming for your life, but that you will now go out and help others through the process of receiving the love of Christ."

Deacon S.L.

Prayer for Restoration:

"When you enter into a relationship and when the relationships that you already have are harmed because of what happened to you, my heart is for you to be completely restored.

I pray that God will bring healing to your marriages and to your relationships with family and friends. I am believing that the incidents of the past will not hinder you. I pray that the weight and the shackle of those incidents will not impact or keep you from your relationships.

May God grant you emotional, spiritual and mental freedom. By the power of God may all that weight be lifted, so that you will have freedom. I pray that there will be a refreshing in your life.

Finally, I pray for you to be able to have full breath in your body and your spirit. In Jesus' name, Amen."

Minister Brian M. Wade

Prayer for Resolution:

Dear God in Heaven, I worship you. You are so amazing and incredible, how could I allow my heart to believe that you are not here? But you are here Lord, you are here with me, and you see me, and you know my situation. In your Word I find that you are a very present help in time of trouble, so I ask you in the name of your holy Son, Jesus Christ, to help me now.

You see this enemy, you know him and you know what he continually comes to do, but Lord, you are stronger than he is, you have already defeated him. So I choose this day to trust you. I will trust you to protect me and keep me. End this menacing, vexing attack on my life. You, Lord are the Alpha and Omega, the Beginning and the End. Therefore I know you can bring an end to this constant advance of the enemy and begin a new hope in me.

Because you are with me, I will not be destroyed. I will not be undone, I will rise up as with the wings of an eagle, soaring high above my trouble on the current of your power. Intervene, oh Lord, help me, save me and heal me.

Because of who You are Lord, I can sleep well and eat well right in the presence of my enemy; I can live. And I will live without fear, anger, tension or stress. I will walk in the spirit by faith, for in Christ Jesus, I am more than a conqueror. Therefore, Lord I celebrate this new day of peace. No longer will I wander around wondering when the next attack may be, but I seek you early to train my hands to war- for you do valiantly oh Lord.

God you are the strength of my life, of whom shall I be afraid? Who is like you Lord? Who is stronger? Who can defeat you? From this moment, I choose to hide under the shadow of your wing, for though I may at times walk in the valley of the shadow of death, I refuse to fear evil because I know that you are with me. Hallelujah! You are with me indeed, and you will never leave me or forsake me.

I also ask that you remind me to forgive perpetually. Every time it comes to mind, I will release it immediately, with your grace. I refuse to hold a grudge or ill will towards anyone. I release forgiveness with all my heart to that one because we all need salvation Lord. Allow the cleansing blood of Jesus, the Lamb of God, to wash this from my heart. I hold all those that harmed me as innocent in my sight, even though I may not be presently able to dwell with them as they are now.

Help them Lord, I know that you are not willing that any should perish; so grant your deliverance and a space to repent. Give them, like you did for King David, a clean heart and a renewed, righteous spirit, so they can start over again.

Thank you, Father, for strengthening my heart and restoring my soul. Thank you for rescuing me. Thank you for giving me hope. This trouble is over; now no one can touch my soul but you. For the earth is yours Lord and the fullness thereof, the world and they that dwell therein. Thank you for hearing and answering our prayers spoken in the mighty, matchless name of Jesus. Amen.

Minister Kesha Lee Gore

"In all their affliction, He was afflicted, and the angel of His presence saved them: in His love and in His pity He redeemed them; and He bare them, and carried them all the days of old." *Isaiah 63:9*

About the Author

Minister Kesha Gore founded SINCERE Ministries, Incorporated in 1989, a year after graduating from Rhema Bible Training Center under the leadership of the late Rev. Kenneth E. Hagin. She has taught hundreds of individuals and groups in churches and conference halls the message of 'Take time to understand the Wisdom of God'. She has authored several books, traveled to South Africa to proclaim the gospel and worked with several ministries and organizations in service to Jesus Christ.

Saved at the age of 12 and called to ministry at the age of 17, Minister Gore learned through suffering, failure and even times of miraculous intervention that God truly is with us. Despite trials, hardness and sheer joy, she has overcome all, through Christ, to continue to extend a genuine, sincere hand to those who truly desire to know the Lord God.

Today, nearly two decades later, she continues to preside over the ministry, write encouraging books and teach people simplistic truths that make it easier to understand the truth of God's Word.

Made in the USA
Lexington, KY
02 January 2012